Endophytic Fungi in Grasses and Woody Plants

Systematics, Ecology, and Evolution

Edited by

Scott C. Redlin
USDA-ARS
Systematic Botany and Mycology Laboratory
Beltsville, Maryland

Lori M. Carris
Washington State University
Pullman

APS PRESS
The American Phytopathological Society
St. Paul, Minnesota

Present address of Scott C. Redlin:
USDA-Animal Plant Health Inspection Service
Plant Protection and Quarantine
Riverdale, MD 20737

Cover: Intercellular hyphae of an endophytic fungus beneath seed coat and extending into the aleurone layer of *Festuca arundinacea* seed. Reprinted, by permission, from *Mycologia* 77:323–327, J. F. White, Jr., and G. T. Cole. Copyright © 1985 The New York Botanical Garden

This book has been reproduced directly from computer-generated copy submitted in final form to APS Press by the editors of the volume. No editing or proofreading has been done by the Press.

Reference in this publication to a trademark, proprietary product, or company name by personnel of the U.S. Department of Agriculture or anyone else is intended for explicit description only and does not imply approval or recommendation to the exclusion of others that may be suitable.

Library of Congress Catalog Card Number: 96-83100
International Standard Book Number: 0-89054-213-9

Printed in the United States of America on acid-free paper

The American Phytopathological Society
3340 Pilot Knob Road
St. Paul, Minnesota 55121-2097, USA

This book is the result of five papers presented at a discussion session of the same title sponsored by the APS Mycology Committee at the annual meeting of The American Phytopathological Society on August 21, 1991, in St. Louis, Missouri. Six additional papers were solicited from scientists and submitted to APS Press. The editors thank David F. Farr, USDA-ARS, Systematic Botany and Mycology Laboratory, Beltsville, Maryland, and Raymond Hammerschmidt, Department of Botany and Plant Pathology, Michigan State University, for assistance in this work.

TABLE OF CONTENTS

Endophytic Fungi in Grasses and Woody Plants

Systematics, Ecology, and Evolution

INTRODUCTION

This book begins with general aspects related to associations between endophytic fungi and their hosts; first a chapter comparing endophytes and latent fungal pathogens, followed by methods for analyzing endophytic communities. A floristic and ecological treatment of fungal endophytes follows in the next seven chapters, starting with the relatively non-specialized group of fungi associated with woody plants and moving to discussions on the more specialized group of Clavicipitaceous and *Acremonium*-like endophytes associated with grasses. The final two chapters address the area of human intervention and endophytic communities: What are the effects of human activities on endophytes, and conversely, can endophytes be manipulated to man's advantage?

An endophyte is literally defined as one plant living within another plant. Endophytic fungi are normally restricted to healthy plant parts and are detected by light microscopy or the use of intense and prolonged surface disinfestation prior to isolating them into pure culture. Two general groups of endophytic fungi are recognized: those that are usually present in grasses, seed transmitted and in the order Clavicipitales or in the genus *Acremonium*; the other group consists of a diverse assemblage of organisms that inhabit woody plants.

The presence of endophytic fungi within plants is incompletely understood but there are a number of ways that they influence plant growth. In grass hosts, they may convey beneficial effects such as the modification of growth habits or the amelioration of the plant's response to drought. They can also result in inconsistent phenotypic expression of disease caused by other organisms. This alteration of disease expression has major implications for resistance breeding programs. Most often they do not cause injury to hosts and some confer resistance in plants to injurious fungi, insects or nematodes. Several are of economic importance because they cause disease in plants: grass choke, caused by *Phomopsis* spp., and some conifer diseases caused by *Rhabdocline parkerii* or *Phyllosticta* spp. *Apiognomonia errabunda* has been reported from the leaves of several woody plant species and exists as an endophyte in *Fagus sylvatica*, occasionally causing anthracnose. Identification of these fungi continues in grasses, ericaceous plants, conifers, broadleaf trees as well as tropical plants such as palms, orchids, aroids and bromeliads.

Many questions about endophytes remain unanswered. What are the factors that cause endophytes to become injurious? Although some may be transmitted in seeds and others stimulate seeds to grow, can they be manipulated to be beneficial in seed propagation? What is their role in compartmentalization of twig dieback and wood decay in trees? Are they effective biocontrol agents? What are their effects on insect herbivory or infection by parasitic fungi or other pathogenic agents?

This new work helps to answer these questions and provide guidance for future research by listing fungal taxa common in various hosts and describing techniques used to isolate and identify the endophytes of Europe, North America and South America that occur in two major groups of plants: grasses and woody plants. Surveys of groups of fungi most frequently encountered, morphological and physiological aspects, coevolution with hosts and adaptive strategies are included. Their adaption to various hosts results in challenges in their study because they may grow slowly in pure culture or sporulate with difficulty or both, resulting in problems in their identification.

Many of the economically important crop plants grown in the United States were introduced from other continents and have not been surveyed for endophytes in their centers of origin or in their more recent regions of production. As plant pathologists, we want to minimize the harmful effects of pathogens and this may be done by optimizing the beneficial effects of endophytes. There is considerable interest in the use of endophytes for biocontrol. This coincides with environmentally conscious demands by the public for sustainable agricultural systems.

CHAPTER 1
LATENT INFECTION VS. ENDOPHYTIC COLONIZATION BY FUNGI

J.B. Sinclair and R.F. Cerkauskas

Department of Plant Pathology
University of Illinois at Urbana-Champaign
Urbana, IL 61801-4739

and

Agriculture Canada Research Station
Vineland Station, Ontario LOR 2EO

The concepts of endophytic colonization and latent infection by fungi are clearly different. All chapters in this book, except this one, will be concerned with the description and study of a variety of endophytic relationships of plants by specialized groups of fungi. This chapter will define and discuss the two concepts, and emphasize latent infection by describing and providing examples in a range of herbaceous and woody dicotyledonous plants. Since there is evidence for latent infection by several fungi in soybean [*Glycine max* (L.) Merr.], it will be used as a case study host to further illustrate latent-infecting fungi in higher plants.

Endophytic Fungi

Endophytic fungi form inconspicuous infections within tissues of healthy plants for all or nearly all their life cycle (100). Endophytes, in contrast to epiphytes, are contained entirely within the substrate plant and may be either parasitic or symbiotic. Endophytic fungi are asymptomatic and may be described as mutualistic (23). The major features of mutualistic symbiosis include the lack of destruction of most cells or tissues, nutrient or chemical cycling between the fungus and host, enhanced longevity and photosynthetic capacity of cells and tissues under the influence of infection, enhanced survival of the fungus, and a tendency toward greater host specificity than seen in nectrophic infections (66). A comparison of the fitness of the host and fungus when living independently in contrast to their fitness when living in association is the major means of determining whether a specific symbiotic association is mutualistic or parasitic (67).

Members of the Ascomycotina, Basidiomycotina, Deuteromycotina, and some Oomycetes have been isolated as endophytes. Endophytic fungi have been isolated from phanerogams in alpine, temperate and tropical regions, although the plants of the Coniferae, Ericaceae and Gramineae have been most intensively sampled (23,86,100).

By definition, endophytic colonization or infection cannot be considered as causing disease, since a plant disease is an interaction between the host, parasite, vector and the environment over time (Fig. 1), which results in the production of disease signs and/or symptoms. The distinction between an endophyte and a pathogen is not always clear. A mutation at a single genetic locus can change a pathogen to nonpathogenic endophytic organism with no effect on host specificity (38). Many pathogens undergo an extensive phase of asymptomatic growth corresponding to colonization and then latent infection before symptoms appear. Many pathogens of economically-important crops may be endophytic or latent in weeds (14,16,46,48,61,70,93). Alternately, nonpathogenic endophytic organisms may play a role as biocontrol agents (38). Both endophytic and latent-infecting fungi can infect plant tissues and become established after penetration, however, infection does not imply the production of visible disease symptoms.

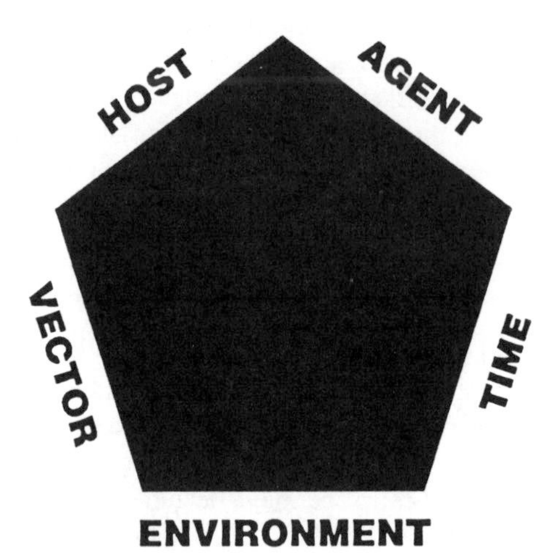

Fig. 1. The five interactive components that can be involved in a symptomatic plant disease.

Latent-Infecting Fungi

Latent-infecting fungi are parasitic but cannot be considered mutualistic. However, some species of *Phomopsis* have been classified as mutualistic endophytes (13). Parasitic fungi subsist in whole or in part upon living tissue and when they cause disease, they are pathogenic. Latent infection of plants by pathogenic fungi has been recognized for many years and is often considered one of the highest levels of parasitism, since the host and parasite coexist for a period of time with minimal damage to the host. True latent infection involves a parasitic relationship that eventually induces macroscopic symptoms (123). Quiescent infections (12), however, are macroscopically visible although mycelial development is arrested after infection and resumes only as the host reaches maturity and/or senescence, while latent contamination involves fungal spores on the host surface which fail to germinate until the host reaches maturity or senescence.

Latent infection is the state in which a host is infected with a pathogen but does not show symptoms and persists until signs or symptoms are prompted to appear by environmental or nutritional conditions or by the state of maturity of the host or pathogen (2). Thus, almost every pathogen, except necrophilic types, has a latent period in which it ramifies host tissues and begins to cause changes in the host's physiology. When biochemical changes become so profound that all or most of the local resources are diverted to the pathogen or if toxic by-products are formed, the affected host tissue becomes symptomatic. One expects that continued conversion of host resources to those of the parasite would eventually result in dwarfing or stunting of the host as one of the first signs of disease.

Latent periods of infections have been described in a variety of host-pathogen interactions. A "latency period" of plant pathogens has been defined as the time required from infection to subsequent production of inoculum (132) and as the time between lesion formation and sporulation (120). These definitions are inadequate because they apply to sporulating fungal pathogens and not to fungal mycelial growth and establishment of nonsporulating bacteria, fungi, mycoplasmalike organisms, nematodes, spiroplasmas, viroids, and those pathogens that cause symptoms before inoculum production. In this chapter, the latent period is considered as the interval from infection to display of macroscopic symptoms, or a "prolonged incubation period". The latent period, which can vary in length of time, usually ends when the plant is under stress or begins to senesce, or is killed by any number of causes (103). Latent infection might be regarded as a type of tolerance of the host to certain pathogens, where the parasite finds the conditions of the internal environment, usually during the early growth stages of the plant unsuitable for completion of its life cycle.

Detection of Latent Infection

Bioassay and histological methods as well as serological techniques can be used to detect latent infection in plant tissues. Bioassay of tissues usually involves the incubation of surface-disinfested seeds or tissues on moist filter paper, blotters or cellulose pads (Kimpac®); or preferably on an agar medium, such as potato-dextrose agar; or a selective medium (33). After a suitable incubation period, usually using ultraviolet or near ultraviolet fluorescent light, the microbial growth is examined using a dissecting microscope for the presence of characteristic fungal fruiting structures. In addition, examination of conidia or other spores using a bright-field compound microscope may be necessary to confirm identification (45,95). Bacteria also may be detected by incubating tissues, preferably on a selective medium, followed by examination of the characteristics of resulting colonies (97,98).

Histopathological studies require thin-sectioning of prepared tissues either by freehand or microtome for those embedded in paraffin or other embedding material, or freeze-dried. Such sections then are studied using a bright-field compound microscope. Both bacteria and fungi can be detected. Differential staining may be required (33).

Serological methods have been developed to detect bacteria and viruses (44) and some fungi in infected plant tissues (121). These include enzyme-linked immunosorbent assay (ELISA), radioimmunoassay (RIA), solid-phase radioimmunoassay (SRIA), and serologically specific electron microscopy (SSEM).

An early technique developed for detection of latent infections by *Colletotrichum* in papaya involved the use of methyl bromide (81). The use of bioassay on moist filter paper, blotters or cellulose pads was enhanced by the discovery that certain desiccant herbicides induce symptoms and fruiting structures of fungi on surface-disinfested, previously asymptomatic tissues. The technique was first developed for use on soybean pod and stem tissues (17,19) and then on young soybean hypocotyl tissues (57). Subsequently, the technique has been used successfully on numerous diverse hosts (Table 1).

The paraquat technique requires that manageable sections of plant tissue be agitated in 70% ethanol for 5 sec, then in 0.5% NaOCl (10% Clorox) for 4 min. (Plant tissues bearing large numbers of soil particles or trichomes may require washing in tap water for 16 to 20 h prior to surface disinfestation.) The tissue then is immersed in sterilized distilled water followed by a dip for 45 to 60 sec in 6000 µg/ml paraquat (1-1'-dimethyl-4,4'-bipyridinium dichloride, Gramoxone 200 g/L). A pre-1984 formulation of the herbicide, paraquat, without sulfacide blue dye or terramins-636 stenching agent is easiest to work with. The paraquat is filter-sterilized and added to sterilized distilled water before use. (CAUTION: Paraquat is fatal if swallowed. Avoid inhalation of most sprays or contact with the eyes or

Table 1. Latent fungal infections in plants other than soybean

Fungus	Host	Literature citations
Alternaria alternata	Persimmon	88
	Mango	89,90
	Tomato	82
Botryosphaeria vaccinii	Cranberry	126
Botrytis spp.	Various hosts	52
Cercospora canescens	Bean*	32
Colletotrichum spp.	Avocado	7
	Banana	77,101,102
	Blueberry	31,47
	Capsicum	1
	Citrus fruit	9,118
	Mango	28,85,101
	Papaya	3,21,34
	Purple passion fruit	55
	*Stylosanthes**	14,65,75,84
	Tomato*	14,93
	Weeds*	14,46,70,93
Cryptosporiopsis curvispora	*Malus*	35
Dothiorella spp.	Mango	54
Guignardia citricarpa	Citrus fruit	71
Lasiodiplodia theobromae	Citrus fruit	10
Leptosphaeria maculans	Canola	41-43,78,131
Monilinia fructicola	Apricot	110,124,125
	Cherry*	130, Cerkauskas (unpublished)
	Peach	110
	Plum*	80,94
Phomopsis citri	Citrus fruit	10
Phomopsis leptostromiformis	Lupin*	25,129
Phomopsis viticola	Grape*	91,92
Phomopsis spp.	Weeds	16,48
Pyricularia grisea	Banana	73

* Paraquat used to determine latent infection.

skin. Follow label directions). Depending on the host and suspect pathogen, treated tissue is incubated for 3 to 4 days under high humidity and continuous fluorescent light at 25 C. When used on soybean plant tissue, mycelia appeared, lesions developed, and fruiting structures of certain fungi developed on surface-disinfested young hypocotyls, stems, leaves and pods (17,19,57). Some paraquat-treated soybean tissues become overgrown by the mycelia of saprophytic fungi during incubation. These mycelia collapse when sprayed lightly with 70% ethanol (69), allowing the fruiting structures of the latent fungi to be studied. However, such a treatment may kill mycelia and spores. A mixture of paraquat plus diquat (25) or diquat alone (J. Northover, Vineland Research Station, Ontario, unpublished), also may be used. Diquat does not have a dye or stenching agent and may be used if the older formulation of paraquat is unavailable. Paraquat also may be incorporated into an agar and fungicide medium (5,83), although a longer period for development of fungal fructifications is required. Glyphosate or a mixture of sodium borate plus sodium chlorate (1:1) also can be used to detect latent fungal infections in soybean tissues (15,72). For example lesions and fruiting structures of *Colletotrichum truncatum* (Schwein.) Andrus & W.T. Moore, and *Macrophomina phaseolina* (Tassi) Goidanich developed 3 wk earlier on soybean plants sprayed with glyphosate than on unsprayed plants (15).

The use of paraquat as a field spray on soybean plants at growth stages R4 to R5 (37) induced symptoms of disease about 2 wk before symptoms appeared on unsprayed plants (15). However, paraquat sprayed on soybean plants at growth stages R6 to R7 reduced seed weight and germination, and increased the incidence of *Alternaria* and *Phomopsis* in seeds of treated plants compared with untreated plants (20).

Paraquat solutions have been used to detect latent infection by fungi in hosts other than soybean (Table 1). For example, a paraquat solution induced symptoms and fruiting structures of *Colletotrichum* spp., including *C. truncatum*, in 17 weeds associated with soybean fields (46). These results suggest either that other plants, in addition to soybean, have developed mechanisms that arrest or delay the progress of microorganisms in their tissues or that latent infection may not be host-specific but rather pathogen-specific. Further study will help resolve this uncertainty.

Treatment of host tissues with paraquat after surface disinfestation causes a loss of membrane integrity with a consequent release of nutrients from plant cells (22). Paraquat generates free radicals in the presence of light, molecular oxygen and photosystem I which result in lipid peroxidation, chlorophyll bleaching, degradation of membranes, and death of the plant (11). Major physiological changes in host tissue during ripening, senescence, or wounding have one common factor, which also results in a decrease in membrane permeability and an increase in release of soluble nutrients (4). In both cases nutrients released from plant cells may affect latency of the fungus by stimulating appressorial germination (76,77,102), or the rapid growth and development of subcuticular hyphae or other infection structures within host tissues. In other instances host factors may induce, maintain or terminate appressorial dormancy because of the close association between fungal development and host senescence (76,77).

Latent Fungal Infections in Herbaceous and Woody Dicotyledonous Plants

Latent infection by diverse species of fungi on various hosts has been reported (Table 1). These hosts include: avocado (*Perisa americana* Miller), blueberry (*Vaccinium corymbosum* L.), banana (*Musa* spp.), Capsicum fruits (*Capsicum* spp.), citrus fruits (*Citrus* spp., *C. sinensis* L.), mango (*Mangifera indica* L.), papaya (*Carica papaya* L.), purple passion fruit (*Passiflora edulis* Sims), soybean [*Glycine max*] (see soybean section), *Stylosanthes* spp., tomato (*Lycopersicon esculentum* L.), and various weeds (53,109).

Studies with *Colletotrichum gloeosporioides* (Penz.) Penz. & Sacc. in Penz. on fruits are not consistent with respect to the survival structures during the latency period. Early observations with tropical fruits, particularly banana, suggested that *Colletotrichum* survived latent periods as subcuticular hyphae which remained quiescent until fruit ripening (102). This also was reported in leaves of citrus, which was colonized by subcuticular hyphae of *Guignardia citricarpa* Kiely, the cause of black spot (71). Meredith (73) reported that *Pyricularia grisea* (Cooke) Sacc. penetrated the cuticle of young banana fruits but not epidermal cell walls and remained latent for months. Currently, dormant appressoria and not subcuticular hyphae are considered important in the latency of *Colletotrichum* on bananas (76,77) and other fruits. For example, in histological studies with field-inoculated attached fruit of avocado and mango, *C. gloeosporioides* survived the latent period as dormant appressoria (7,28). Dormant subcuticular hyphae were not observed. Similar observations were made with detached *Capsicum annuum* L. fruits inoculated with *Glomerella cingulata* (Stoneman) Spauld. & H. Schrenk and *Colletotrichum capsici* (Syd.) E. J. Butler & Bisby (1). In detached mature green papaya fruit, however, the ultrastructural evidence showed that *C. gloeosporioides* formed appressoria with infection pegs and subcuticular hyphae with the latter structure possibly serving as a dormant form of the fungus (21,34). Further detailed studies are required to confirm this view. In contrast, *C. gloeosporioides* persisted as germinated appressoria and penetration hyphae in attached blueberry fruit which were susceptible to infection at all stages of development (31). In attached seedling leaves of summer orange (*Citrus natsudaidai* Hayata), the fungus persisted during the latent period as vesiculate hyphae occupying the intercellular space of the leaf epidermis or was closely attached to the outside or inside of the epidermal cell membrane. The hyphae originated from the infection thread which emerged from the appressorium (118). In detached mature citrus fruit, however, *C. gloeosporioides* persisted on the surface as latent appressoria, the majority of which did not produce infection hyphae, and those hyphae that were formed resulted in latent infection before fruit maturity. The latent hyphae of the fungus were present within and beneath the cuticle and in the intercellular spaces of the epidermis. The formation of infection hyphae by appressoria was affected by relative humidity. By fruit maturity latent infection occurred within three or four cell layers of the flavedo (9). The infection hyphae were similar to those formed by *Cryptosporiopsis curvispora* (Peck) Gremmen in Boerema & Gremmen during the penetration of the cuticle and cortical cells of apples (35).

Low levels of latent colonization of *Stylosanthes guianensis* (Aubl.) Sw. var. *guianensis* by *C. gloeosporioides* (14) and the teleomorph, *Glomerella cingulata* (84), have been reported. We note that increasing anthracnose severity has been observed with advancing plant maturity (29,30,51). An indirect association with frequency of defoliation (75) and large diurnal temperature fluctuations (65) also may be factors.

Latent infection by *Colletotrichum coccodes* (Wallr.) S. J. Hughes occurs in tomato foliage and fruit (39,68) and weeds (14,93). The fungus, which ramifies the tissue, originates from an appressorium, remains latent in green tomato fruits, and survives as infection hyphae between the cell wall and cutin layer. As fruits ripen the fungus produces inter- and intracellular hyphae culminating in the formation of lesions (39,68). However, no studies on the survival of *C. coccodes* structures during latency in weeds have been reported.

Latent and quiescent infection by fungi other than *Colletotrichum* also are well known particularly those caused by species of *Botrytis* such as *B. aclada* Fresen., *B. cinerea* Pers.:Fr., *B. convoluta* Whetzel & Drayton, *B. elliptica* (Berk.) Cooke, *B. fabae* Sardiñã, *B. narcissicola* Kleb. and *B. tulipae* (Lib.) Lind. *B. cinerea* has the largest host range and includes leaves of rye (*Secale cereale* L.) and strawberry (*Fragaria*); flowers of apple (*Malus*), black currant (*Ribes americanum* L.), eggplant (*Solanum melongena* L.), grape (*Vitis*), pear (*Pyrus communis* L.), potato (*Solanum tuberosum* L.), raspberry (*Rubus*), rose (*Rosa*) and strawberry; and fruits of tomato, as well as other host tissues. These fungi survive the latent period as inactive hyphae within host tissue (52). Agar medium supplemented with chloramphenicol and paraquat were used to study infection and colonization by indirect estimation of the sporulation potential by *B. cinerea* on leaf disks, petals and stamens of strawberry (83).

Latent infection by *Alternaria alternata* (Fr.:Fr.) Keissl. of various fruits has been reported (Table 1). In these reports the mycelia remain in a latent state within the fruit throughout the growing season until ripening or tissue senescence when fruits gradually lose their resistance, whereupon the fungus renews its development. On persimmon (*Diospyros kaki* L.) and mango fruits germinating conidia directly penetrate the fruit cuticle and lenticels, respectively, and the hyphae remain latent in the intercellular spaces. Upon maturity, intercellular darkening and cell collapse occur and the fungus resumes growth in the intercellular spaces (88,90). Infections of green tomato fruits by *A. alternata* produced tiny quiescent lesions in which the fungal mycelium was located intercellularly in the epidermis and subepidermal tissue. Necrosis did not extend beyond the third cell layer of the lesion (82).

Botryosphaeria vaccinii (Shear) Barr establishes latent infections in cranberry (*Vaccinium macrocarpon* Aiton) fruits and leaves. A localized superficial blemish may occur in dormant fruit infections although further symptom development is delayed until after harvest. However, latent infections of leaves by dormant hyphae within the cuticle or between the cuticle and outer epidermal wall cannot be detected macroscopically and remain dormant until abscission occurs (127).

Isolation of *Dothiorella* from mango stems indicated that latent infection may occur and the stem end of fruits are colonized during maturation by hyphae (54). Similarly, stem-end rot, a major source of decay of citrus fruits, caused by *Lasiodiplodia theobromae* (Pat.) Griffon & Maubl. and *Phomopsis citri* H. Fawc. non (Sacc.) Traverso & Spessa, hom. illeg. arises from latent infections in the stem button (calyx plus disk) of the fruit. Although infection may occur at any stage prior to harvest under favorable environmental conditions, disease development does not occur until the fruit is harvested. At this time the stem bottom becomes senescent and begins to separate from the fruit. Neither fungus enters healthy tissues of the button until separation of the cuticle and epidermis in the abscission zone. At this point extensive intra- and intercellular hyphal growth occurs through the albedo of the fruit (10).

Various species of *Phomopsis* have been associated with latent infection in a variety of plants. *Phomopsis leptostromiformis* (Kühn) Bubàk in Kab. & Bubàk has a long latent period in narrow-leaf lupins and does not normally cause symptoms on living plants. However, distinctive subcuticular coralloid latent hyphae form in lupin stems infected with the fungus (129). Paraquat was used to detect and quantify the extent of this latent infection. This may allow for screening of lupins for resistance to Phomopsis stem blight, in detecting variations in relative virulence in pathogen populations, and determining environmental conditions suitable for infection (25).

The paraquat technique was used to study the epidemiology of *Phomopsis viticola* (Sacc.) Sacc., cause of cane and leaf spot of grape. Infections due to *P. viticola* at bloom remain latent since fruit rot symptoms usually do not develop until close to harvest. Latent fruit and rachis infections were detected using the technique although histological studies of this phase of the disease are lacking (91,92).

In a different application of the paraquat technique, latent infection by seedborne *Cercospora canescens* Ellis & G. Martin, cause of leaf spot and blotch of bean (*Phaseolus vulgaris* L.), was detected by spraying seedlings with the herbicide (32). The fungus formed colonies on the upper side of the herbicide-sprayed cotyledons. The technique was also used to study *Monilinia fructicola* (G. Wint.) Honey infections in immature fruit of plum (*Prunus*) (80) and cherry (*Prunus*) (130, R.F. Cerkauskas unpublished) where latent and quiescent infections were reported (94). These also were reported in apricot (*Prunus armenica* L.) (110,124,125) and peaches (*Prunus persica* (L.) Batsch) (110). In plum fruit, infection by hyphal penetration of stomata and lenticels occurred at any time during fruit development (118) while in cherry fruit penetration occurred through stomata or directly through the cuticle (26). Invasion of fruits was arrested shortly after infection and resumed upon the initiation of fruit ripening. Histological studies showed that after stomatal infection in immature apricot fruit between full-bloom and shuck-fall stages further growth by the fungus ceased until the fruit began to ripen, when disease symptoms appeared (123,124).

Latent infection by *Leptosphaeria maculans* (Desmaz.) Ces. & De Not., the cause of blackleg of canola or rape (*Brassica napus* L. and *B. rapa* L.) is important in the epidemiology of the disease. Infection of stems in the field may precede the appearance of symptoms by several months (131) and older plants have a longer latent period than younger plants (41). Five stages by which the fungus enters the stem via the leaf and petiole have been described (42,43). These consist of symptomless colonization of the leaf, followed by leaf necrosis and systemic symptomless colonization of the petiole and stem, and finally, by stem necrosis. The systemic symptomless colonization of the stem occurs via an extensive network of hyphae ramifying intercellularly in the cortical cell layers near the epidermis (78). These cells are identical to cells in areas lacking intercellular hyphae and in epidermal peels from uninoculated plants (78). The intercellular hyphae are also common within the lacunae and deeper within the cortex (78). The hyphae grow intercellularly in the petiole within the cortex or within the xylem vessels without macroscopic symptoms (42,43).

Most yield losses occur from crown cankers which develop primarily by infection of young plants from the cotyledon to the eighth leaf stage through this systemic route of symptomless colonization. Crown infection may precede expression of canker symptoms by several weeks during the growing season and symptom progression may not occur during the winter season (41). Most infections via this systemic pathway occurred during the early susceptible stage of plant development (131). However, a long period of latent infection and symptom expression are closely related to the level of host resistance, whereas the establishment of early season infection is not linked. Thus, both latent infection and visible symptoms need to be considered in the identification and use of resistant genotypes particularly in view of the high frequency of latent infection observed in resistant cultivars (131).

Latent Fungal Infections in Soybeans - A Case Study

Latent infection in soybeans is the colonization or infection of tissues by fungal pathogens for a prolonged period without visible symptoms. Soybeans have developed mechanisms, not yet understood, that arrest or delay the progress of many microorganisms reaching internal tissue during the plant's vegetative and early reproductive stages. Latent infection in soybeans, in our opinion, results from a process of coevolution between the crop and pathogens that allows the ultimate accumulation of many resistance and fitness genes in the host and various parasites. Highly susceptible individual plants and highly virulent pathogens are eliminated early in the coevolutionary process.

Most fungal and viral pathogens of soybean seedlings, plants, pods and seeds have an asymptomatic or latent period after infection. Less research has been done on latent infection by bacterial, nematode and viral, pathogens than by fungal pathogens. Of more than 100 organisms known to infect soybeans only about 35 are economically important (103). Of these, at least 10 to 15, including bacteria and fungi, cause disease in which an extended latent period separates pathogen introduction and symptom expression. The latent period may be several weeks. No definitive studies have been done on the effect of latent infection by any pathogen on soybean plant development, growth or yield.

All who work with soybeans in the field have observed that plants severely damaged or killed by an abiotic factor may show symptoms or signs of disease caused by a biotic agent. Such plant damage is often attributed to the pathogen inducing the most conspicuous symptoms, when actually other factors may have caused plant destruction or death. For example, soybean plants severely stressed or killed any time during the growing season may show symptoms of anthracnose caused by *C. truncatum* or charcoal rot caused by *M. phaseolina*, since both fungi infect the plants and roots early in the season and remain latent until infected plants become stressed.

Soybean plants sprayed with the fungicide benomyl in the field tend to remain green longer than unsprayed plants (87,104). This phenomenon, attributed to delay in plant senescence, is largely due to the fungistatic activity of benomyl on fungi causing latent infection and delayed symptom development on aging or stressed plants. Enhancement of yield and seed quality has been attributed to the use of benomyl sprays when soybean plants are stressed or harvest is delayed. These observations provide indirect evidence that latent infection by fungi does adversely affect soybeans. However, the actual effects of pathogens on infected tissues or on the plant during the latent period are virtually undetermined, especially in soybeans.

At least 14 fungal pathogens cause latent infection in soybeans (Table 2). Six of these will be discussed in detail: those that infect aboveground plant parts - *Cercospora kikuchii* (Matsumoto & Tomoyasu) M. W. Gardner, cause of leaf spot and bronzing, and purple seed stain; *Cercospora sojina* K. Hara., cause of frogeye leaf spot; *C. truncatum* cause of anthracnose; members of the *Diaporthe/Phomopsis* complex [*Diaporthe phaseolorum* (Cke. & Ell.) Sacc. var. *sojae* (S. G. Lehman) Wehmeyer (anamorph *Phomopsis phaseoli* (Desmaz.) Sacc., cause of pod and stem blight; *Diaporthe phaseolorum* var. *caulivora* Athow & Caldwell (anamorph rare), cause of northern stem canker; *Diaporthe phaseolorum* f. sp. *meridionalis* (anamorph unknown) Morgan-Jones, cause of southern stem canker; and *Phomopsis longicolla* T. W. Hobbs, cause of Phomopsis seed decay]; and *M. phaseolina*, cause of charcoal rot; and those that infect roots - *C. truncatum*, *M. phaseolina* and *Phytophthora sojae* M. J. Kauffman & J. W. Gerdemann, cause of Phytophthora root rot.

Latent infection in aboveground tissues

C. kikuchii usually induces foliage symptoms at the time of seed set (growth stages R3 to R4) regardless of environmental conditions. Symptom development is associated with physiological changes in the plant during transition from vegetative to reproductive stages. Sources of inoculum are infected seeds or infested surface debris (103). Seedlings emerging from infected seeds may become infected during emergence, whereas older plants may become infected by spores produced on crop debris throughout the growing season. Neither infected seedlings nor young plants may show symptoms. For example, symptomless pods and stems of two soybean cultivars previously inoculated with a spore suspension of *C. kikuchii* in the field developed more lesions with stromata and conidia of the fungus when tissues were treated with paraquat than when tissues were not treated (17,19), thereby implicating latent infection.

C. truncatum, cause of anthracnose, has a latent period in soybeans (116,117) and other *Colletotrichum* spp. can establish latent infections in many other hosts (Table 1) (14,46,70,117). Soybean plants are susceptible to *Colletotrichum* at all stages of development, but symptoms on above ground tissue typically appear in the early reproductive stages (growth stages R1 to R2). Severe symptoms develop after prolonged periods of high humidity, as plants senesce, or when they become stressed. The source of primary inoculum is infected soybean seeds, infested crop debris, and infected alternative hosts (104). Latent infection by *C. truncatum* in soybeans was demonstrated when greater numbers of acervuli appeared 3 wk earlier on field-grown plants sprayed with paraquat than on unsprayed ones (15,17). Latent infections were also detected in field-grown soybean leaves and petioles and freshly cut stubble dipped in paraquat but not on untreated tissues (46,47).

Table 2. Fungi with a latent period in soybean plants and seeds

Fungus	Possible latent period according to growth stages[a]	Literature citations
Cercospora kikuchii	V1 to R4-R5	17,19,105
C. sojina	V1 to R4-R5	8
Colletotrichum destructivum (teleomorph: *Glomerella glycines*)	V1 to R7	14,15
C. gloeosporioides (teleomorph: *Glomerella cingulata*)	Unknown	14
C. truncatum	V1 to R1-R7	17,18,46,64
Diaporthe phaseolorum var. *caulivora*	V1 to R6-R7	60
D.p. f. sp. *meridionales*	V1 to R6-R7	107
D.p. var. *sojae* (anamorph. *Phomopsis phaseoli*)	V1 to R6-R7	8,15,59
Fusarium	Unknown	17
F. oxysporum	Unknown	122
Macrophomina phaseolina	V1 to R7	50,63,69
Phialophora gregata	V4 to R2-R3	99
Phomopsis longicolla	Unknown	36,49,59
Phytophthora sojae	V2 to R5-R6	74,106

[a] Time of infection and environmental factors influence length of latent periods. Growth stages from Fehr *et al.* 1971 [Crop Sci. 11:929-931 (37)].

Gerdemann (39) was first to suggest that *D. phaseolorum* var. *sojae* was latent in soybean tissues. Latent infection by members of the *Diaporthe/Phomopsis* complex in soybeans was implicated in studies of systemic fungicides to control seedborne pathogens of soybeans (36,86,107). Evidence for latent infection by *Phomopsis* was shown when the fungus was recovered from symptomless stems and senescent cotyledons of 4-wk-old soybean plants (59), and evidence for latent infection by *P. phaseoli* was shown when this fungus was recovered from various symptomless parts of soybean plants in the vegetative stage (59,61). Additionally, *Phomopsis* was isolated from 12-day-old symptomless seedlings and green pods and *D.*

phaseolorum var. *sojae* was isolated from 30- to 33-day-old symptomless plants (60). When full green (growth stage R6) and yellow (growth stage R7) soybean pods were inoculated with *P. phaseoli*, lesions developed in only 5 and 26% of the pods, respectively, but the fungus was isolated from all inoculated pods (49). Bioassays of other maturing soybean tissues without symptoms recovered *D.p. caulivora*, cause of northern soybean stem canker (61). In addition, *Phomopsis* colonized the vascular system of inoculated soybean plants in the mid- to late-vegetative stages (growth stages V6-V8) without showing symptoms (61). Treatment of symptomless pods and stems from field-grown plants with paraquat elicited rapid formation of pycnidia of *P. sojae* and allowed for rapid detection of extent of latent colonization by the fungus (17,19). Also, *P. longicolla* and *P. phaseoli* were detected early in asymptomatic soybean tissue using polyclonal antibodies developed against *P. longicolla* (121).

Charcoal rot usually appears on soybean plants after midseason after a period of drought or when plants reach senescence. Symptoms appeared late in the season even though 80 to 100% of the seedlings in the field may be infected 2 to 3 wk after planting (104). Bioassay and histological methods can be used to detect the fungus in asymptomatic soybean seedlings early in the growing season and to show microsclerotia in the vascular system (50). Latent infection by *M. phaseolina* was established in soybeans when the number of pycnidia on soybean stems was greater after paraquat treatment than by visual rating in the field (69).

Latent infection in roots

Six isolates of *C. truncatum* were pathogenic to soybean roots (57). Symptoms and fruiting structures were induced on seedling roots of plants by placing excised roots on soybean yeast-extract agar for 3 to 4 days at 27 C and on hypocotyls from the same seedlings only after the latter were treated with paraquat (56). In addition to *C. truncatum* and *M. phaseolina*, latent root infection by fungal pathogens that primarily attack the above ground parts of soybean plants has been shown for *P. sojae* (74,104,106), cause of Phytophthora root and stem rot. The fungus is usually evident late in the growing season (104). Plant height and yield of infected plants are lower than those of uninfected plants, despite the absence of stem browning and lesion development in the greenhouse or field (74). Also, oospores are found in the roots of asymptomatic seedlings of cv. Amsoy 71, tolerant of *P. sojae* (106). Soybean seedlings may serve as a latent source of pathogen inoculum for the mature-plant phase of the disease over a wide temperature range (106).

Latent infection in seeds

Infected or infested soybean seeds transmit pathogens and other microorganisms. The terms "infected" or "internally seedborne" and "infested" or "externally seedborne" refer to the location of the microorganism in or on the seed. Soybean seeds infested with most microorganisms would appear symptomless, except when a microorganism grows extensively over the surface. An example is the encrustation by surface growth of the downy mildew fungus, *Peronospora manshurica* (Naumov) Syd. in Gäum, on soybean seeds. Generally, the expression of colonization in soybean seeds by most pathogens is exceptional.

The bioassay, histological and serological techniques used to detect microorganisms and viruses in soybean plant tissues can be used for seeds. However, treatment with a desiccant or other herbicide does not enhance detection of latent infection of soybean seeds by microorganisms. Bioassays have recovered the following fungal genera from surface-disinfested soybean seeds (104): *Acremonium, Alternaria, Aspergillus, Botrytis, Cercospora, Chaetomium, Choanephora, Cladosporium, Diplodia, Fusarium, Penicillium, Pestalotia, Pythium, Rhizopus, Sclerotinia,* and *Thielavia.* A few of these fungi induce symptoms. A complete list of characteristic fungi associated with soybeans and soybean seeds has been published (103).

Histological techniques have shown the following pathogenic fungi to colonize and infect soybean seeds: *A. alternata* (62), *A. tenuissima* (Kunze:Fr.) Wilshire (104), *Cercospora kikuchii* (105), *C. sojina* (64), *Colletotrichum truncatum* (64), members of the *Diaporthe/Phomopsis* complex (58,104), *F. oxysporum* Schlechtend.:Fr. (122), and *M. phaseolina* (63). Although these fungi may induce symptoms on seeds, they are generally latent.

Latent-infecting fungi in soybean seed coats can be identified without staining based on color and width of mycelia, formation of hyphal aggregates, mycelial mats or sclerotia, and the presence or absence of oil globules in the hyphae (Table 3) (64,105,122).

Epidemiological and control implications

Epidemiological. Although evidence has accumulated on latent infection by various fungal pathogens of soybeans, the role of latent infection in the epidemiology of these pathogens requires further investigation (14,15). An early infection that does not result in conspicuous signs or symptoms may weaken the plant, predisposing it to other stresses or diseases, or may even kill it (i.e. damping-off). The epidemiology of plant diseases is partially based on the visual development and spread of disease in a plant population, on a single plant or on plant parts. The establishment and detection of latent infection would influence the recognition of disease spread within a population. Limited stress areas within a population would result in symptom development on some plants and none on others, even though most plants might have latent infection. Soybean anthracnose, charcoal rot, and Phytophthora root and stem rot are good examples. Little is known about the stress placed on soybean plants by latent-infection microorganisms, either alone or in combination. Our experience has been that soybean stem pieces from field-grown plants will develop symptoms and fruiting structures of more than one fungus under laboratory conditions. These infecting fungi require a food base during a latent period, but whether such infection has other than a minor effect on plant growth and eventual yield is not known. This is not a symbiotic relationship. We suspect there may be measurable adverse effects. Although some information exists on the effects of various environmental factors on latent infection, more can be learned from continued studies.

Latent infection generally restricts host colonization for a longer period and thus reduces potential inoculum production. Perhaps the ability of fungi to infect soybeans without symptoms is important for the survival of these fungi (14,15). By restricting its domination in host tissues, the pathogen allows for the continued growth and reproduction of the plant. As the plant becomes senescent, for whatever reason and at whatever time, the fungus dominates the physiology of the tissues, symptoms are induced, and sporulation occurs. Most fungi that are latent in soybean plants are also seedborne, except for *Phialophora gregata, Phytophthora sojae*, and *M. phaseolina* (50,99). These three fungi are primarily soilborne in crop debris, and *M. phaseolina* can be seedborne (104). Inoculum production on plants as seeds mature would increase the possibilities of the fungi to be established in seed tissues. Ramification in tissues and development of overseasoning structures, such as microsclerotia and oospores, would establish the soilborne fungi in crop debris.

Disease control. Latent infection is important in the epidemiology and control of plant disease and also in the breeding for resistance or tolerance to a pathogen. An understanding of latent infection contributes to development of effective control measures, as does an understanding of penetration, colonization, disease expression and yield losses.

Knowledge of the latent phase of any pathogen, the length of that latency, and the mechanisms that trigger the pathogen to induce symptoms and to reproduce is important in the improvement of control measures. If pathogen infection occurs early in the season without symptoms, control measures taken later in the season could be less effective or ineffective. Control might require less effort and expenditure during the latent period infection, when the crop plant is small and less tissue is compromised, than after symptoms appear. For example, the prediction of Phomopsis seed decay of soybeans by detection of latent infection of *P. longicolla* in pods resulted in determining the need and timing for a fungicide spray to control the disease in Iowa (72). Also, the detection of latent fungal infection in soybean seedlings by use of paraquat was important in developing a model relating temperature, duration of surface wetness, and rain to seedling infection (96).

Latent infection might be regarded as a type of tolerance or resistance to

Table 3. Appearance of various fungi in soybean seed coat tissues in thin section prepared for bright-field microscopy[a]

Fungus	Hyphal width (μm)	Unstained mature hyphae	Hyphal aggregate	Mycelial mat	Sclerotia	Oil globules
Alternaria alternata	1.8-5.4	Brown	No	Yes	No	No
Cercospora kikuchii	1.3-3.0	Light brown	Yes	No	No	No
C. sojina	1.3-2.7	Brown	Yes	Yes	No	No
Colletotrichum truncatum	3-11	Brown	No	Yes	No	Yes[b]
Fusarium	1.4-3.6	Hyaline	No	No	No	No
Macrophomina phaseolina	1.8-4.5	Brown	No	No	Yes	No
Phomopsis	3.8-8.7	Hyaline	Yes	Yes	No	Yes

[a] Data from Kunwar *et al.,* 1985 (64), Singh and Sinclair, 1985 (105), and Velicheti and Sinclair, 1991 (122).

[b] Prominent oil globules

certain pathogens, where the parasite finds the internal environment unsuitable for growth and multiplication. Such resistance or tolerance prevents rapid multiplication of microorganisms that reach the plant interior (79). The mechanisms involved in latent infection, the inheritance of such mechanisms, and the influence of the environment on the expression of these mechanisms can be important in developing resistant or tolerant cultivars (24,131).

Latent infection has been referred to as nonrace-specific or horizontal-type resistance, because such resistance is effective to some degree against more or all races of the pathogen involved (79). The degree of this type of resistance is directly correlated with the length of the latency period (79). Selecting for disease resistance among soybean cultivars and lines to one of the latent fungi can become complicated because of latent infection. Selection for resistance is based on symptom development. If symptom development by latent pathogens is triggered by plant senescence or stress, disease evaluation can be difficult. For example, in a segregating population, the development of aboveground symptoms may be related more to host susceptibility to an environmental factor or root damage than to the latent-infection fungus.

Weed control

Many pathogenic fungi, particularly species of *Colletotrichum*, have been considered as potential mycoherbicides to control weeds (111,113-115). Tests conducted for host specificity and host range are important factors in the selection of mycoherbicides (126). Before a potential mycoherbicide is used commercially, however, latent infection by the fungus in target and nontarget plants should be ascertained to reduce the risk of early sporulation of organisms that might be found later to be pathogenic to important agricultural crops (14). Potential mycoherbicides generally are tested by spray application of spore inoculum onto host and nonhost plants in the laboratory and field. Plants are then monitored for symptom development for 2 to 4 wk after inoculation but not until maturity and tests for latent infection are not included (27,112,128). The development of Collego®, a mycoherbicide containing *C. gloeosporioides* f. sp. *aeschynomene* provides such an example (108). This mycoherbicide was registered for control of northern jointvetch (*Aeschynomene virginica* (L.) B.S.P.) in soybean fields in Arkansas, Mississippi, and Louisiana yet testing for latent infection by the fungus in soybean and other plants was not performed prior to registration, patent application, or commercial release (27,108). Additional pathogenicity tests based on visible symptom expression indicated that only species within the subfamily Papilionoidea were susceptible to infection (112); however, tests for latent infection were not performed. Subsequent tests using paraquat demonstrated extensive recovery of the fungus from symptomless soybean and various other economic plants and weeds that were inoculated earlier with the fungus (14), in contrast to findings by TeBeest (112) who reported many of these species as immune to infection on the basis of visible symptoms only. Also, in laboratory and greenhouse tests, *C. g.* f. sp. *aeschnomene* reduced soybean seed germination and radicle length and caused some stunting of seedlings in comparison to controls (14). Furthermore, application of the fungus on flowering fields during the recommended period may result in latent infection (14) and may cause increases in overwintering inocula especially in fields where paraquat has been used as a harvest aid (15,20).

The overseasoning of *Colletotrichum* on soybean stubble and weeds (46), the latent infection by these and other fungi of soybean and various weeds (14,46,59-61,116) as well as the lack of host specificity of diverse *Colletotrichum* and *Phomopsis* isolates from weeds and economic plants (6,14,46,61,103) indicate the necessity for comprehensive testing of specificity and host range of facultative saprophytes proposed as mycoherbicides for use on soybean or other economic plants. (14). The paraquat technique may facilitate such testing by reducing time spent waiting for plants to reach maturity (14).

LITERATURE CITED

1. Adikaram, N.K.B., Brown, A.E., and Swinburne, T.R. 1983. Observations on infection of *Capsicum annum* fruit by *Glomerella cingulata* and *Colletotrichum capsici*. Trans. Br. Mycol. Soc. 80:395-401.
2. Agrios, G.N. 1988. Plant Pathology. 3rd ed. Academic Press, New York. 803 pp.
3. Alvarez, A.M., and Nishijima, W.T. 1987. Postharvest diseases of papaya. Plant Dis. 71:681-686.
4. Baddeley, M.S. 1971. Biochemical aspects of senescence. Pages 415-429 in: Ecology of Leaf Surface Micro-organisms. T.F. Preece and C.H. Dickinson, eds. Academic Press, London. 641 pp.
5. Bannon, E. 1978. A method of detecting *Septoria nodorum* on symptomless leaves of wheat. Irish J. Agric. Res. 17:323-325.
6. Batson, W.E., and Roy, K.W. 1982. Species of *Colletotrichum* and *Glomerella* pathogenic to tomato fruit. Plant Dis. 66:1153-1155.
7. Binyamini, N., and Schiffmann-Nadal, M. 1972. Latent infection in avocado fruit due to *Colletotrichum gloeosporioides*. Phytopathology 62:592-594.
8. Bisht, V.S., and Sinclair, J.B. 1985. Effect of *Cercospora sojina* and *Phomopsis sojae* alone or in combination on seed quality and yield of soybeans. Plant Dis. 69:436-439.
9. Brown, G.E. 1975. Factors affecting postharvest development of *Colletotrichum gloeosporioides* in citrus fruit. Phytopathology 65:404-409.
10. Brown, G.E., and Wilson, W.C. 1968. Mode of entry of *Diplodia natalensis* and *Phomopsis citri* into Florida oranges. Phytopathology 58:736-739.
11. Burden, R.S., Cooke, D.T., and Hargreaves, J.A. 1990. Mechanism of action of herbicidal and fungicidal compounds on cell membranes. Pestic. Sci. 30:125-140.
12. Byrde, R.J.W., and Willetts, H.J. 1977. The Brown Rot Fungi of Fruit: Their Biology and Control. Pergamon Press, New York. 171 pp.
13. Carroll, G.C. 1986. The biology of endophytism in plants with particularreference to woody perennials. Pages 205-222 in: Microbiology of the Phyllosphere. N.J. Fokkema and J. van den Heuvel, eds. Cambridge Univ. Press, Cambridge. 392 pp.
14. Cerkauskas, R.F. 1988. Latent colonization by *Colletotrichum* spp.: Epidemiological considerations and implications for mycoherbicides. Can. J. Plant Pathol. 10:297-310.
15. Cerkauskas, R. F., Dhingra, O. D., and Sinclair, J. B. 1983. Effect of three desiccant-type herbicides on fruiting structures of *Colletotrichum truncatum* and *Phomopsis* spp. on soybean stems. Plant Dis. 67:620-622.

16. Cerkauskas, R.F., Dhingra, O.D., Sinclair, J.B., and Asmus, G. 1983. *Amaranthus spinosus, Leonotis nepetaefolia*, and *Leonurus sibiricus*: New hosts of *Phomopsis* spp. in Brazil. Plant Dis. 67:821-824.
17. Cerkauskas, R.F., and Sinclair, J.B. 1980. Use of paraquat to aid detection of fungi in soybean tissues. Phytopathology 70:1036-1038.
18. Cerkauskas, R.F., and Sinclair, J.B. 1982. Effect of paraquat on soybean pathogens and tissues. Trans. Br. Mycol. Soc. 78:495-502.
19. Cerkauskas, R.R., Sinclair, J.B., and Floor, S.R., 1979. The use of paraquat to detect latent colonization by fungi of soybean stem and pod tissues. (Abstr.) Page 85 in: Proc. World Soybean Res. Conf. 2nd. F. T. Corbin, ed. Westview Press, Boulder, CO.
20. Cerkauskas, R.F., Dhingra, O.D., Sinclair, J.B., and Foor, S.R. 1982. Effect of three dessicant herbicides on soybean (*Glycine max*) seed quality. Weed Sci. 30:484-490.
21. Chau, K.F., and Alvarez, A.M. 1983. A histological study of anthracnose on *Carica papaya*. Phytopathology 73:1113-1116.
22. Chia, L.S., Thompson, J.E., and Dumbroff, E.B. 1981. Simulation of the effects of leaf senescence on membranes by treatment with paraquat. Plant Physiol. 67:415-420.
23. Clay, K. 1991. Fungal endophytes, grasses, and herbivores. Pages 199-252 in: Microbial Mediation of Plant of Plant-Herbivore Interactions. P. Barbosa, V.A. Krischik, and C.G. Jones, eds. John Wiley & Sons, Inc., New York.
24. Cowling, W.A., Hamblin, J., Wood, P. McR., and Gladstone, J.S. 1987. Resistance to Phomopsis stem blight in *Lupinus angustifolius* L. Crop Sci. 27:648-652.
25. Cowling, W.A., Wood, P. McR., and Brown, A.G.P. 1984. Use of paraquat-diquat herbicide for the detection of *Phomopsis leptostromiformis* infection in lupins. Australasian Plant Pathol. 13:45-46.
26. Curtis, K.M. 1928. The morphological aspect of resistance to brown spot in stone fruit. Ann. Bot. 42:39-68.
27. Daniel, J.T., Templeton, G.E., Smith Jr., R.J., and Fox, W.T. 1973. Biological control of northern jointvetch in rice with an endemic fungal disease. Weed Sci. 21:303-307.
28. Daquioag, V.R., and Quimio, T.H. 1978. Latent infection in mango caused by *Colletotrichum gloeosporioides*. Philipp. Phytopathol. 15:35-46.
29. Davis, R.D., Irwin, J.A.G., Cameron, D.F., and Shepherd, R.K. 1987. Epidemiological studies on the anthracnose diseases of *Stylosanthes* spp. caused by *Colletotrichum gloeosporioides* in North Queensland and pathogenic specialization within the natural fungal populations. Aust. J. Agric. Res. 38:1019-1032.

30. Davis, R.D., Irwin, J.A.G., Shepherd, R.K., and Cameron, D.F. 1987. Yield losses caused by *Colletotrichum gloeosporioides* in three species of *Stylosanthes*. Aust. J. Exp. Agric. 27:67-72.
31. Daykin, M.E., and Milholland, R.D. 1984. Infection of blueberry fruit by *Colletotrichum gloeosporioides*. Plant Dis. 68:948-950.
32. Dhingra, O.D., and Asmus, G.L. 1983. An efficient method of detecting *Cercospora canescens* in bean seed. Trans. Br. Mycol. Soc. 81:425-426.
33. Dhingra, O.D., and Sinclair, J.B. 1985. Basic Plant Pathology Methods. CRC Press, Inc., Boca Raton, FL. 355 pp.
34. Dickman, M.B., and Alvarez, A.M. 1983. Latent infection of papaya caused by *Colletotrichum gloeosporioides*. Plant Dis. 67:748-750.
35. Edney, K.L. 1958. Observations on the infection of Cox's Orange Pippin apples by *Gloeosporium perennans* Zeller & Childs. Ann. Appl. Biol. 46:622-629.
36. Ellis, M.A., Machado, C.C., Prasartsee, C., and Sinclair, J.B. 1974. Occurrence of *Diaporthe phaseolorum* var. *sojae* (*Phomopsis* sp.) in various soybean seedlots. Plant Dis. Rep. 58:173-176.
37. Fehr, W.R., Caviness, C.E., Burmood, D.T., and Pennington, J.S. 1971. Stages of development descriptions for soybean (*Glycine max* L.) Merr. Crop Sci. 11:929-931.
38. Freeman, S., and Rodriguez, R.J. 1993. Genetic conversion of a fungal plant pathogen to a nonpathogenic, endophytic mutualist. Science 260:75.
39. Fulton, J.P. 1948. Infection of tomato fruits by *Colletotrichum phomoides*. Phytopathology 38:235-245.
40. Gerdemann, J.W. 1954. The association of *Diaporthe phaseolorum* var. *sojae* with root and basal stem rot of soybean. Plant Dis. Rep. 38:742-743.
41. Hall, R. 1992. Epidemiology of blackleg of oilseed rape. Can. J. Plant Pathol. 14:46-55.
42. Hammond, K.E., Lewis, B.G., and Musa, T.M. 1985. A systemic pathway in the infection of oilseed rape plants by *Leptosphaeria maculans*. Plant Pathol. 34:557-565.
43. Hammond, K.E., and Lewis, B.G. 1987. The establishment of systemic infection in leaves of oilseed rape by *Leptosphaeria maculans*. Plant Pathol. 36:135-147.
44. Hampton, R., Ball, E., and De Boer, S. 1990. Serological Methods for Detection and Identification of Viral and Bacterial Plant Pathogens: A Laboratory Manual. American Phytopathological Society, St. Paul, MN. 389 pp.
45. Hanlin, R.T. 1990. Illustrated Genera of Ascomycetes. American Phytopathological Society, St. Paul, MN. 263 pp.

46. Hartman, G.L., Manandhar, J.B., and Sinclair, J.B. 1986. Incidence of *Colletotrichum* spp. on soybeans and weeds in Illinois and pathogenicity of *Colletotrichum truncatum*. Plant Dis. 70:780-782.
47. Hartung, J.S., Burton, C.L., and Ramsdell, D.C. 1981. Epidemiological studies of blueberry anthracnose disease caused by *Colletotrichum gloeosporioides*. Phytopathology 71:449-453.
48. Hepperly, P.R., Kirkpatrick, B.L., and Sinclair, J.B. 1980. *Abutilon theophrasti*: Wild host for three fungal parasites of soybean. Phytopathology 70:307-310.
49. Hepperly, P.R., and Sinclair, J.B. 1980. Detached pods for studies of *Phomopsis sojae* pods and seed colonization. J. Agric. Univ. P.R. 64:330-337.
50. Ilyas, M.B., and Sinclair, J.B. 1974. Effects of plant age upon development of necrosis and occurrence of intraxylem sclerotia in soybean infected with *Macrophomina phaseolina*. Phytopathology 64:156-157.
51. Irwin, J.A.G., Cameron, D.F., and Lenne, J.M. 1984. Responses of *Stylosanthes* to anthracnose. Pages 295-310 in: The Biology and Agronomy of *Stylosanthes*. H.M. Stace and L.A. Edye, eds. Academic Press, Sydney. 656 pp.
52. Jarvis, W.R. 1977. *Botryotinia* and *Botrytis* Species: Taxonomy, Physiology, and Pathogenicity, A Guide to the Literature. Can. Dept. Agric. Monograph 15. 195 pp.
53. Jeffries, P., Dodd, J.C., Jeger, M.J., and Plumbley, R.A. 1990. The biology and control of *Colletotrichum* species on tropical fruit crops. Plant Pathol. 39:343-366.
54. Johnson, G.I., Mead, A.J., Cooke, A.W., and Dean, J.R. 1991. Mango stem end rot pathogens - Infection levels between flowering and harvest. Ann. Appl. Biol. 119:465-473.
55. Kagiwata, T. 1986. An anthracnose of passion fruit caused by *Glomerella cingulata* (Stoneman) Spaulding et Schrenk. J. Agric. Sci., Japan. 3:90-100.
56. Khan, M., and Sinclair, J.B. 1991. Effect of soil temperature on infection of soybean roots by sclerotia-forming isolates of *Colletotrichum truncatum*. Plant Dis. 75:1282-1285.
57. Khan, M., and Sinclair, J.B. 1992. Pathogenicity of sclerotia- and nonsclerotia-forming isolates of *Colletotrichum truncatum* on soybean plants and roots. Phytopathology 82:314-319.
58. Kilpatrick, R.A., and Johnson, H.W. 1953. Fungi isolated from soybean plants at Stoneville, Mississippi, in 1951-1952. Plant Dis. Rep. 37:98-100.
59. Kmetz, K., Ellett, C.W., and Schmitthenner, A.G. 1974. Isolation of seedborne *Diaporthe phaseolorum* and *Phomopsis* from immature soybean plants. Plant Dis. Rep. 58:978-982.

60. Kmetz, K.T., Schmitthenner, A.F., and Ellett, C.W. 1978. Soybean seed decay: Prevalence of infection and symptom expression caused by *Phomopsis* sp., *Diaporthe phaseolorum* var. *sojae*, and *D. phaseolorum* var. *caulivora*. Phytopathology 68:836-840.
61. Kulik, M.M. 1984. Symptomless infection, persistence, and production of pycnidia in host and non-host plants by *Phomopsis batatae*, *Phomopsis phaseoli* and *Phomopsis sojae*, and the taxonomic implications. Mycologia 76:274-291.
62. Kunwar, I.K., Manandhar, J.B., and Sinclair, J.B. 1986. Histopathology of soybean seeds infected with *Alternaria alternata*. Phytopathology 76:543-546.
63. Kunwar, I.K., Singh, T., Machado, C.C., and Sinclair, J.B. 1986. Histopathology of soybean seed and seedling infection by *Macrophomina phaseolina*. Phytopathology 76:532-535.
64. Kunwar, I.K., Singh, T., and Sinclair, J.B. 1985. Histopathology of mixed infections by *Colletotrichum truncatum* and *Phomopsis* spp. or *Cercospora sojina* in soybean seeds. Phytopathology 75:489-492.
65. Lenne, J.M. 1985. Recent advances in the understanding of anthracnose of *Stylosanthes* in tropical America. Pages 773-775 in: Proc. 15th Int. Grassland Congr. Kyoto, Japan.
66. Lewis, D.H. 1973. Concept in fungal nutrition and the origin of biotrophy. Biol. Rev. 48:261-278.
67. Lewis, D.H. 1974. Micro-organisms and plants: The evolution of parasitism and mutualism. Pages 367-392 in: 24th Symposium of the Society for General Microbiology. M.J. Carlile, and J.J. Skehel, eds. Cambridge University Press, Cambridge, U.K.
68. Ludwig, R.A. 1960. Host pathogen relationships with the tomato anthracnose disease. Pages 37-56 in: Proc. Plant Science Seminar, Campbell Soup Co., Camden.
69. Machado, C.C., Hartman, G.L., Manandhar, J.B., and Sinclair, J.B. 1984. Detection of *Macrophomina phaseolina* in soybean stems. (Abstr.) Prog. and Abstracts, World Soybean Res. Conf. III:46. Iowa State Univ. Press, Ames.
70. McLean, K.S., and Roy, K.W. 1988. Incidence of *Colletotrichum dematium* on prickly sida, spotted spurge, and smooth pigweed and pathogenicity to soybean. Plant Dis. 72:390-393.
71. McOnie, K.C. 1967. Germination and infection of citrus by ascospores of *Guignardia citricarpa* in relation to control of black spot. Phytopathology 57:743-746.
72. McGee, D.C. 1986. Prediction of Phomopsis seed decay by measuring soybean pod infection. Plant Dis. 70:329-333.
73. Meredith, D.S. 1963. *Pyricularia grisea* (Cooke) Sacc. causing pitting disease of bananas in Central America. Ann. Appl. Biol. 52:453-463.

74. Meyer, W.A., and Sinclair, J.B. 1972. Root reduction and stem lesion development on soybeans by *Phytophthora megasperma* var. *sojae*. Phytopathology 62:1414-1416.
75. Miles, J.W., and Lenne, J.M. 1987. Effect of frequency of defoliation of 40 *Stylosanthes guianenis* genotypes on field reaction to anthracnose caused by *Colletotrichum gloeosporioides*. Aust. J. Agric. Res. 38:309-315.
76. Muirhead, I.F. 1981. The role of appressorial dormancy in latent infection. Pages 155-167 in: Microbial Ecology of the Phylloplane. J.P. Blakeman, ed. Academic Press, New York. 502 pp.
77. Muirhead, I.F., and Deverall, B.J. 1981. Role of appressoria in latent infection of banana fruits by *Colletotrichum musae*. Physiol. Plant Pathol. 19:77-84.
78. Nathaniels, N.Q.R., and Taylor, G.S. 1983. Latent infection of winter oilseed rape by *Leptosphaeria maculans*. Plant Pathol. 32:23-31.
79. Nelson, R.R. 1979. The evolution of parasitic fitness. Pages 23-46 in: Plant Disease. An Advanced Treatise. Vol. 4. How Pathogens Induce Disease. J.G. Horsfall and E.B. Cowling, eds. Academic Press, New York.
80. Northover, J., and Cerkauskas, R.F. 1992. Detection and control of *Monilinia fructicola* latent infections in plums. (Abstr.) Phytopathology 82:1069.
81. Parris, G.K., and Jones, W.W. 1941. The use of methyl bromide as a means of detecting latent infections by *Colletotrichum* spp. Phytopathology 31:570-571.
82. Pearson, R.C., and Hall, D.H. 1975. Factors affecting the occurrence and severity of blackmold of ripe tomato fruit caused by *Alternaria alternata*. Phytopathology 65:1352-1359.
83. Peng, G., and Sutton, J.C. 1991. Evaluation of microorganisms for biocontrol of *Botrytis cinerea* in strawberry. Can. J. Plant. Pathol. 13:247-257.
84. Pereira, J.O., and Azevedo, J.L. 1993. Endophytic fungi of *Stylosanthes*: A first report. Mycologia 85:362-364.
85. Peterson, R.A. 1986. Mango Diseases. Pages 233-247 in: Proc. of the CSIRO First Australian Mango Research Workshop, 1984, Cairns, Australia. 392 pp.
86. Petrini, O. 1986. Taxonomy of endophytic fungi of aerial plant tissues. Pages 175-187 in: Microbiology of the Phyllosphere. N.J. Fokkema and J. van den Heuvel, eds. Cambridge University Press, Cambridge. 392 pp.
87. Prasartsee, C., Tenne, F.D., Ilyas, M.B., Ellis, M.A., and Sinclair, J.B. 1975. Reduction of internally seedborne *Diaporthe phaseolorum* var. *sojae* by fungicide sprays. Plant Dis. Rep. 59:20-23.

88. Prusky, D., Ben-Arie, R., and Guelfat-Reich, S. 1981. Etiology and histology of Alternaria rot of persimmon fruits. Phytopathology 71:1124-1128.
89. Pursky, D., Fuchs, Y., and Zauberman, G. 1981. A method for pre-harvest assessment of latent infections in fruits. Ann. Appl. Biol. 98:79-85.
90. Prusky, D., Fuchs, Y., and Yanko, U. 1983. Assessment of latent infections as a basis for control of postharvest disease of mango. Plant Dis. 67:816-818.
91. Pscheidt, J.W., and Pearson, R.C. 1989. Effect of grapevine training systems and pruning practices on occurrence of Phomopsis cane and leaf spot. Plant Dis. 73:825-828.
92. Pscheidt, J.W., and Pearson, R.C. 1989. Time of infection and control of Phomopsis fruit rot of grape. Plant Dis. 73:829-833.
93. Raid, R.N., and Pennypacker, S.P. 1987. Weeds as hosts for *Colletotrichum coccodes*. Plant Dis. 71:643-646.
94. Rosenberger, D.A. 1985. Observations on quiescent brown rot infections in Grand Prize plums. Pages 19-22 in: Proc. Brown Rot of Stone Fruit Workshop: Ames, Iowa, July 11, 1983., Special Rept. New York Agric. Exp. Sta., Geneva, New York.
95. Rossman, A.Y., Palm, M.E., and Spielman, L.J. 1987. A Literature Guide for the Identification of Plant Pathogenic Fungi. American Phytopathological Society, St. Paul, MN. 252 pp.
96. Rupe, J.C., and Ferriss, R.S. 1987. A model for predicting the effects of microclimate on infection by soybean by *Phomopsis longicolla*. Phytopathology 77:1162-1166.
97. Saettler, A.W., Schaad, N.W., and Roth, D.A., eds. 1989. Detection of Bacteria in Seed and Other Planting Material. American Phytopathological Society, St. Paul, MN. 122 pp.
98. Schaad, N.W., ed. 1988. Laboratory Guide for Identification of Plant Pathogenic Bacteria. 2nd ed. American Phytopathological Society, St. Paul, MN. 164 pp.
99. Schneider, R.W., Sinclair, J.B., and Gray, L.E. 1972. Etiology of *Cephalosporium gregatum* in soybean. Phytopathology 62:345-349.
100. Siegal, M.R., Latch, G.C.M., and Johnson, M.C. 1987. Fungal endophytes of grasses. Annu. Rev. Phytopathol. 25:293-315.
101. Simmonds, J.H. 1941. Latent infection in tropical fruits discussed in relation to the part played by species of *Gloeosporium* and *Colletotrichum*. Proc. Roy. Soc. Qld. 52:92-120.
102. Simmonds, J.H. 1963. Studies in the latent phase of *Colletotrichum* species, concerning ripe rots of tropical fruits. Queensl. J. Agr. Sci. 20:373-424.
103. Sinclair, J.B. 1991. Latent infection of soybean plants and seeds by fungi. Plant Dis. 75:220-224.

104. Sinclair, J.B., and Backman, P.A., eds. 1989. Compendium of Soybean Diseases. 3rd ed. American Phytopathological Society, St. Paul, MN. 106 pp.
105. Singh, T., and Sinclair, J.B. 1985. Further studies on the colonization of soybean seeds by *Cercospora kikuchii* and *Phomopsis* sp. Seed Sci. Technol. 14:71-77.
106. Slusher, R.L., and Sinclair, J.B. 1973. Development of *Phytophthora megasperma* var. *sojae* in soybean roots. Phytopathology 63:1165-1171.
107. Smith, E.F., and Backman, P.A. 1989. Epidemiology of soybean stem canker in the southeastern United States: Relationship between time of exposure to inoculum and disease severity. Plant Dis. 73:464-468.
108. Smith Jr., R.J. 1986. Biological control of northern jointvetch (*Aeschynomene virginica*) in rice (*Oryza sativa*) and soybeans (*Glycine max*) - A researcher's view. Weed Sci. 34:17-23.
109. Swinburne, T.R. 1983. Quiescent infections in post-harvest diseases. Pages 1-21 in: Post-harvest Pathology of Fruits and Vegetables. C. Dennis, ed. Academic Press Inc., London. 264 pp.
110. Tate, K.G., and Corbin, J.B. 1978. Quiescent fruit infections of peach, apricot, and plum in New Zealand caused by the brown rot fungus *Sclerotinia fructicola*. N.Z. J. Exp. Agric. 6:319-325.
111. TeBeest, D.O., Yang, X.B., and Cisar, C.R. 1992. The status of biological control of weeds with fungal pathogens. Annu. Rev. Phytopathol. 30:637-657.
112. TeBeest, D.O. 1988. Additions to host range of *Colletotrichum gloeosporioides* f. sp. *aeschynomene*. Plant Dis. 72:16-18.
113. TeBeest, D.O., and Templeton, G.E. 1985. Mycoherbicides: Progress in the biological control of weeds. Plant Dis. 69:6-10.
114. Templeton, G.E. 1982. Biological herbicides: discovery, development, deployment. Weed Sci. 30:430-433.
115. Templeton, G.E., and TeBeest, D.O. 1979. Biological weed control with mycoherbicides. Annu. Rev. Phytopathol. 17:301-310.
116. Tiffany, L.H. 1951. Delayed sporulation of *Colletotrichum* on soybean. Phytopathology 41:975-985.
117. Tiffany, L.H., and Gilman, J.C. 1954. Species of *Colletotrichum* from legumes. Mycologia 46:52-75.
118. Tokunaga, Y., and Ohira, I. 1973. Latent infection of anthracnose on citrus in Japan. Rept. Tottori Mycol. Inst. (Japan) 10:693-702.
119. Valleau, W.D. 1915. Varietal resistance of plums to brown-rot J. Agric. Res. 5:365-396.
120. Van der Plank, J.E. 1963. Plant Diseases: Epidemics and Control. Academic Press, New York. 349 pp.
121. Velicheti, R.K., Lamison, C., Brill, L.M., and Sinclair, J.B. 1993. Immunodetection of *Phomopsis* spp. in asymptomatic soybean plants. Plant Dis. 77:70-73.

122. Velicheti, R.K., and Sinclair, J.B. 1991. Histopathology of soyabean seeds colonized by *Fusarium oxysporum*. Seed Sci. Technol. 19:445-450.
123. Verhoeff, K. 1974. Latent infections by fungi. Annu. Rev. Phytopathol. 12:99-107.
124. Wade, G.C. 1956. Investigations on brown rot of apricots caused by *Sclerotinia fructicola* (Wint.) Rehm. 1. The occurrence of latent infection in fruit. Aust. J. Agric. Res. 7:504-515.
125. Wade, G.C., and Cruickshank, R.H. 1992. The establishment and structure of latent infection with *Monilinia fructicola* in apricots. J. Phytopathol. 136:95-106.
126. Watson, A.K. 1984. Host specificity of plant pathogens in biological weed control. Pages 577-586 in: Proc. VI Int. Symp. Biol. Contr. Weeds, Vancouver, Canada. 885 pp.
127. Weidemann, G.J., and Boone, D.M. 1984. Development of latent infections on cranberry leaves inoculated with *Botryosphaeria vaccinii*. Phytopathology 74:1041-1043.
128. Weidemann, G.J., and D.O. TeBeest. 1990. Biology of host range testing for biocontrol of weeds. Weed Tech. 4:465-470.
129. Williamson, P.M., Sivasithamparam, K., and Cowling, W.A. 1991. Formation of subcuticular coralloid hyphae by *Phomopsis leptostromiformis* upon latent infection of narrow-leafed lupins. Plant Dis. 75:1023-1026.
130. Wittig, H.P., Johnson, K.B., and Pscheidt, J.W. 1991. Potential of resident epiphytic fungi for biological control of brown rot blossom blight in stone fruits. (Abstr.) Phytopathology 81:1153.
131. Xi, K., Morrall, R.A.A., Gugel, R.K., and Verma, P.R. 1991. Latent infection in relation to the epidemiology of blackleg of spring rapeseed. Can J. Plant Pathol. 13:321-331.
132. Zadoks, J.C. 1978. Methodology of epidemiological research. Pages 63-96 in: Plant Disease. An Advanced Treatise. Vol. 2. How Disease Develops in Populations. J.G. Horsfall and E.B. Cowling, eds. Academic Press, New York. 436 pp.

CHAPTER 2

ISOLATION AND ANALYSIS OF ENDOPHYTIC FUNGAL COMMUNITIES FROM WOODY PLANTS

Gerald F. Bills

Microbial Biochemistry and Process Research
Merck Research Laboratories
P.O. Box 2000
Rahway, NJ 07065

The past 15 years has seen a proliferation of investigations that are only beginning to uncover the vast internal mycota of living plant tissues and its functions. For the purpose of this discussion, a broad definition of endophytism is used, that being fungi established inside healthy plant tissue without causing overt symptoms in or apparent injury to the host (54, 99). Although unequivocal proof of their concealed internal occupation of plants by fungi should be obtained by direct microscopic observation, correlative histological studies verifying internal colonization patterns of endophytes have been neglected (23). Growth of hyphae outwardly from internal tissues of rigorously surface-sterilized plant organs generally is considered evidence that these fungi are endophytic during routine screening of plant organs. This expanded definition of endophytes subsumes many fungi historically called latent and facultative pathogens, portions of the life cycles of many common epiphytic fungi, plus an incredible variety of other fungi with at least limited capacity to persist inside living plant tissue. The term "endophyte", whose common usage has evolved only in recent times, is now accepted as a useful name for categorizing a myriad of plant-associated fungi familiar to those engaged in diagnosis of plant diseases and identification of plant-inhabiting fungi. Many of the fungi dominant in endophytic communities are species familiar to forest pathologists and their mention reemphasizes the vagueness of the definitions dividing endophytes, facultative pathogens, and latent pathogens.

Soon after phytopathologists understood that fungi growing internally in plants caused disease, they attempted to isolate these organisms into axenic culture to determine what were the casual agents of disease and to reintroduce these presumptive disease agents into host tissues to satisfy the conditions of Koch's postulates. The isolation of fungi from woody plants to identify and localize latent pathogens has been practiced since the beginning of the 20th century (40, 116, 117, 138).

Long before the recent interest in endophytic fungal communities, basic methods for study of endophytes had been well-established by those studying fungal diseases of woody plants and interactions among fungal pathogens and other fungi resident within the same host. What differs about ongoing investigations of endophytic fungi, is that single causal organisms of disease have not been pursued, but all plant tissues have been surveyed on a systematic basis for culturable internal fungi in a manner analogous to ecological investigations of vegetation landscapes. During studies of various host plants, the internal fungi have been sampled in very methodical patterns with the goal of describing the spatial distributions and temporal sequences of these fungal assemblages. Investigators have tried to ascertain what are the underlying anatomical features, developmental attributes, host preferences, environmental and climatic cues and geographical factors that determine colonization strategies of endophytic fungi (99). Because of the inherent obstacles to direct observation and identification of fungi inside living plants, probing endophytic fungal communities has depended heavily on observations and correlative data compiled from exhaustive isolations from plant tissues. As always, the interpretation of microbial community structure based on isolations of the organisms produces biased results related to the selective action of the isolation methods and sampling patterns.

Most procedures for isolating endophytes are comparatively simple and routine for one skilled in basic plant pathological or microbiological technique. However, the process of designing an analytical study of an endophyte community, handling and maintaining the often hundreds of isolates, characterizing the isolates taxonomically, and quantitatively interpreting the results can be burdensome and overwhelming. The techniques and materials used for isolation, maintenance, identification and preservation of endophytes of grasses were reviewed recently (6). Grass endophytes are an extremely host-specific and specialized subset of endophytes with their own peculiar life cycles. The isolation and identification methods for endophytes of non-grass phanerogams was last reviewed by Petrini (97) in which he presented a useful table listing surface sterilization protocols for various kinds of plants and plant organs.

Since that time, dozens of articles have appeared that describe the isolations and analyses of fungal communities inhabiting woody plants. A staggering accumulation of data is now available that correlates the occurrence of fungi within organs of various plant taxa (19). A discussion of various isolation techniques, sampling strategies, and identification methods for endophytes of leaves, stems, and roots of woody plants is presented below.

Intersection of Endophytes with Other Fungal Communities

After reviewing lists of fungi isolated from woody plant organs, one may be asking "what fungi are not endophytic?" The core group of organisms consists of ascomycetous fungi and their conidial states (97). Many of the genera are familiar to forest pathologists and mycologists as the causal agents of leaf-spots, needle casts, anthracnoses, die-backs, or cankers. These fungi are also encountered as secondary wound invaders, epiphytes, wood-decay basidiomycetes, soft-rot fungi, soil fungi, insect pathogens, coprophilous fungi, or aquatic hyphomycetes. Thus part of the fascination surrounding endophyte research stems from the variety of interrelationships found among different kinds of fungal communities and the attempt to understand the reasons that drive all these organisms to congregate quiescently inside vascular plants.

Epiphytic saprobes associated with the phylloplane are among the most frequently isolated endophytes (23, 54). Soil fungi can be isolated at a low frequency from leaves, but with higher frequency from bark (16, 37) and are the dominant fungi isolated from roots (55, 56, 68). Basidiomycetes are rarely detected as endophytes in leaves or stems of small shrubs, but have been recovered from tree bark and sapwood (16, 33, 64) particularly if benomyl is used as a selective agent in the media. Insect pathogens, e.g. *Beauveria bassiana* (Bals.) Vuill., *Paecilomyces farinosus* (Holmsk.) Brown & G. Smith and *Verticillium lecanii* (A. Zimmerm.) Viegas, have been recovered from living bark (16) suggesting that bark may serve as a natural reservoir for saprobic existence of these fungi (25, 26). Isolation studies of grasshopper gut contents and fecal pellets demonstrated that grasshoppers grazing tropical vegetation are capable of ingesting and dispersing viable propagules of endophytes (83). Research on dispersal by herbivorous insects, especially those with strict host preferences or those capable of simultaneous feeding on both native vegetation and crops, merits further study to assess the potential of insects as vectors of endophytic fungi.

Endophytic fungi inhabiting leaf and stem tissue in the canopy of coniferous forests can dominate early stages of litter decomposition (5, 74, 81, 82). The role of endophytism as a mechanism to capture resources during early colonization and decomposition of angiosperm wood is now starting to be understood (19, 33, 63). Successional studies of deciduous (144) and evergreen (114) angiosperm trees have determined the fate of epiphytes and endophytes after senescent leaves are deposited in the litter. Many of the dominant genera obtained from living leaves, e.g. *Xylaria, Nodulisporium, Geniculosporium, Idriella, Fusarium, Cylindrocarpon, Hormonema, Exophiala, Phomopsis, Ascochyta, Colletotrichum, Cryptosporiopsis, Sphaeropsis*, are readily isolated from fresh angiosperm forest litter (Bills, Polishook & Peláez, unpublished). Thus endophyte assemblages form a spatial and temporal continuum through forest strata, beginning with latent invaders of the canopy and extending to those entering the upper litter horizons that are eventually replaced in the lower litter layers by secondarily saprobic fungi.

Viable propagules of endophytic fungi can be recovered in great numbers from lichen thalli (103). During an intensive examination of only 17 fruticose lichen samples, each consisting of 1 g fresh weight, 506 different taxa of fungi were recovered, of these 306 were isolated only once. This outstanding level of species recovery was achieved by simultaneous application of serial washing of minute lichen fragments, selective weeding of unwanted colonies, a medium containing a potent but non-killing colony-restricting agent, and temperature manipulation (discussed below). A specific group of lichenicolous symbionts could not be identified, but many genera and species generally considered plant endophytes and phytopathogens were present.

Aquatic hyphomycetes, fungi specifically adapted to sporulate and disperse while submerged under water, are a common component isolated from the interior of riparian tree roots that are at least periodically submerged in fresh water streams (52, 56, 124, 125). Numerous aquatic hyphomycete species can persist both in living xylem and bark of roots in these habitats, as well as numerous soil fungi that are typical of terrestrial roots. However, separate analysis of root bark and xylem demonstrated greater incidence of fungi in the bark and that this is the site of primary invasion by aquatic hyphomycetes, with invasion of the xylem occurring subsequently. Another similar group of fungi with highly modified conidia adapted for water dispersal are termed "terrestrial aquatic hyphomycetes" (2). They sporulate from surfaces of living leaves or stems exposed to dew droplets, fog, or rain. To observe and isolate conidia of these fungi, water droplets from leaf or stem surfaces are collected in plastic bags. The liquid is gently centifuged and the resultant sediment is either fixed for microscopic observation or spread onto isolation media for selection of single conidium isolates (3).

In view of the expanded definition of endophytes and the recognition that many kinds of fungi form benign or mutualistic association in roots, the distinctions between endophytes and certain types of mycorrhizae, e.g. ectendomycorrhizae, ericoid mycorrhizae, and pseudomycorrhizae become indistinct. Many dematiaceous fungi, some identified as *Mycelium radicis atrovirens* Melin and the form genera *Phialophora*, *Phialocephala*, and *Chloridium* are associated with roots of conifers (68, 139). Certain mutalistic root-inhabiting or mycorrhizal fungi associated with plants in the Ericaceae (39, 129) and Orchidaceae (38, 140) have been referred to as endophytes.

Ascomycetous coprophilous fungi are isolated consistently but with low frequency from leaves and stems of woody plants (97). These fungi often possess ascospores with thickened cell walls and gelatinous sheaths or appendages. The ascomata are adapted to launch their spores onto the caulo- or phylloplane. A triangular life cycle has evolved where spores are ejected from dung, adhere to vegetation, and then disperse to new locations after herbivorous animals ingest the plants and pass them through their digestive system. The fact that studies of terrestrial or aquatic root endophytes have failed to isolate coprophilous forms is consistent with their adaption for dispersal to aerial plant parts.

Collection and Isolation Strategies

The needs of the investigator will necessarily dictate which approach will be followed for plant selection and isolation procedures for endophytes. Analyses of large numbers of isolates from even limited sampling can strain personnel and laboratory facilities, often resulting in little more information than a checklist of species. Therefore it is imperative to choose realistic methods and appropriate experimental design before hand.

If isolates of a particular latent pathogen or endophyte that commonly fruits on the host are desired, then it may be just as simple to wait until the host dies, the fungus sporulates on the surface, and then propagate isolates from spores in fertile conidiomata or ascomata rather than searching through the dozens of species that can be isolated from the host. Likewise isolation from lesions or fruiting bodies is useful for establishing the cultural identity or conspecificity of two different morphological states of the same organism. This is also the most effective method for obtaining haploid isolates necessary for genetic analysis of population structure. The mechanics and utility of direct isolation of fungi from conidia and ascospores have been described in detail (73).

Dominant endophytes may prevail throughout specific tissues and are convenient organisms for autoecological and genetic studies of spatial distribution. Patterns of mycelial density and of spatial extent of individuals will influence isolation frequency scores; therefore sampling patterns ideally should reflect the actual spatial extent of the individual mycelia. Sampling scale may be guided by inspection of recently dead host material for developing decay zones or stained or discolored areas in the wood, or from sapwood adjacent to fruiting bodies. Patterns of fruiting body formation may also provide clues to the extent of underlying mycelia. Methods for mapping and genetic analyses of mycelial individuals of decay fungi in the underlying sapwood and xylem are now well-established (106) and are applicable to latent fungi in bark and xylem. If mycelial distributions are not obvious, then leaf or stem fragments or small chips of wood can be removed along longitudinal or transverse axes of stems at prescribed intervals. Fragments then are plated in serial order and the relative origin of isolates are recorded. Intraspecific pairing of mycelial isolates with subsequent evaluation of somatic or mating compatibility reactions can reveal spatial extent of similar or dissimilar genotypes.

Serial microdissection in combination with light microscopy of individual cleared needles of *Pseudotsuga menziesii* (Mirab.) Franco led to the recognition that *Rhabdocline parkeri* Sherwood, Stone & Carroll occurred as multiple individual infections in needles and was localized in hypodermal and epidermal cells (128). Serial microdissection has revealed a similar colonization pattern for *Lophodermium piceae* (Fuckel) Höhn. in needles of *Picea abies* Karst. Two dimensional maps of multi-species infections of *Eucalyptus* leaves were constructed by plating 3 mm^2 leaf squares in serial order (13).

Ordinations of dominant endophyte communities from specific organs of a host taken along strong environmental gradients, such as altitude, canopy height, degree of canopy closure, site moisture, and wind exposures have been useful in discerning abiotic factors that influence endophyte infection (27, 68, 98, 104). For example, regression analysis positively correlated snowfall and endophytic isolation frequency from red alder twigs (*Alnus rubra* Bong.), but not bigleaf maple (*Acer macrophyllum* Pursh) (120). Endophyte infection was lower in leaves of the *Loiseleuria procumbens* (L.) Desv. in alpine tundra sites with high wind exposure (98). Few studies have examined the geographic influences on distribution of endophytes. Occasionally floristic lists of fungi isolated from the same host at distant sites have been compared (24, 53). But to date, relative densities of endophytes or endophyte species richness in relation to plant diversity along latitudinal gradients have not been examined experimentally. Do equivalent plant organs, under equivalent moisture regimes, harbor equivalent numbers of endophytes in arctic, temperate, and tropical biomes?

Culturing different age classes of conifer needles separately has led to the recognition that frequency of colonization of needles generally increases with age (94). In leaves of *Sequoia sempervirens* (D. Don) Endl., species diversity increases up to a plateau at 3-7 years and then declines as *Cryptosporiopsis abietina* Petrak and *Pleuroplaconema* sp. become dominant in the oldest leaves (48). Mapping frequency scores of multiple species revealed small scale patchy distributions in leaves of increasing age sampled along single branches, but without clear demarcation of a successional pattern. Direct counts of infections by *Rhabdocline parkeri* follow a T log model with increasing needle age (128).

Tissue and organ specificity, or lack of, among endophytes can be demonstrated through careful dissection and separate culturing and analysis of adjacent tissues or organs and these kinds of experiments have been performed extensively (26, 99). A classic example of simultaneously testing hypotheses about host and tissue specificity was carried out in a mixed stand of *Pinus sylvestris* L. and *Fagus sylvatica* L. (102). Whole stem sections and decorticated stem sections of each tree species were cultured separately. Many of the most frequently isolated fungi exhibited host specificity even though the two hosts were next to each other. Occasionally taxa dominant on one host, e.g. *Verticicladium trifida* Preuss on *P. sylvestris*, a species generally associated with conifers, was observed to crossover to the other host and colonize a similar niche on *Fagus*. Several fungi preferentially colonized the bark, while being absent in the underlying xylem tissues. Cluster and discriminant analysis of fungal frequency scores indicated that each of the 4 sample types, whole or decorticated stems of each host, were distinguishable by its component fungi thus supporting the view that most endophytes exhibited both host and tissue specificity. Coprophilous species were preferentially isolated from bark of *P. sylvestris* when compared to *F. sylvatica*, perhaps because the thick-walled ascospores lodged in cracks of the rough pine bark could evade surface sterilization. Occurrence of specialized xylotropic endophytes (31) deep within whole tree stems was demonstrated by sawing whole *Picea abies* stems (55-60 yr old) (110-112) first transversely and then each transverse cylinder longitudinally. Cultures were made from small cubes of wood removed aseptically along transverse axes at each level above the soil. Highest frequency scores and species diversity occurred in the outer 0-2 cm. Some basidiomycete species and *Nectria fuckeliana* Booth were consistently isolated up to 6-7 cm deep. However, recovery of *Ascocoryne* spp. and *Neobulgaria premnophila* Roll-Hansen & Roll-Hansen increased in frequency with greater depth into the stems.

"Do endophytes and other fungi influence the establishment and spread of bark and stem pathogens?" has been a question of interest in forest pathology. Establishment and spread of virulent and hypovirulent strains of chestnut blight (*Cryphonectria parasitica* (Murrill) Barr) were monitored in stem-girdled *Quercus rubra* L. and *Castanea dentata* (Marsh.) Bork (7). Establishment and spread of *C. parasitica* strains inoculated into stems of both species were measured by extracting 5 mm bark plugs at prescribed distances from the inoculation point months after inoculation. Bark plugs were surface sterilized and incubated on a glucose-extract medium and frequencies of culturable fungi enumerated. Not only could *C. parasitica* be reisolated from a high percentage of the trees, but also a large number of genera of endophytic species were found to coexist in the bark with all strains of *C. parasitica.* Thus it was concluded that the normal bark mycota did not influence establishment and spread of hypovirulent strains. Cotter and Blanchard (37) surveyed the microflora of living bark of *Fagus grandifolia* Ehrh. to determine if differences in microbial composition could be correlated with cankering of bark and tree exposures. To isolate fungi, 2 mm diameter bark chips were extracted from stems in the field and placed directly on isolation plates of beech extract agar, benomyl malt agar, or malt yeast agar. Except for the beech bark disease pathogen, *Nectria coccinea* (Pers.) Fr. var. *faginata* Lohman, Watson & Ayers, similar assemblages of endophytic and epiphytic fungi were recovered from both healthy and cankered bark and from all compass points of the trees.

Shigo (119) compared the succession of fungi between internal tissues of untreated oaks (upland species of *Quercus* subgenus *Erythrobalanus*) infected by oak wilt (*Ceratocystis fagacearum* (Bretz) J. Hunt) and infected oaks that were deep-girdled to the heartwood to limit maturation and sporulation of the pathogen. Small wood chips were removed from tangential slabs from trunks, and limb and root sections. Chips were flame-sterilized with ethanol and planted in malt-yeast agar. Besides recovering *C. fagacearum* as the most frequently isolated fungus, typical endophytes, *Hypoxylon punctulatum* (Berk. & Rav.) M. C. Cooke, *Hypoxylon atropunctatum* and *Botryodiplodia gallae* (Schwein.) Petrak & Sydow (= *Dothiorella quercina* (M. C. Cooke & Ellis) Sacc.) were recovered from a high percentage of the deep-girdled trees. It is now known that endophytically established fungi, such as *Hypoxylon* spp., undergo active mycelial development in response to water stress in host organs (19, 31). The reduced vigor and dissemination of *C. fagacearum* in deep-girdled trees may well have been caused by its displacement by endophytes that rapidly colonized the oak sapwood.

Distribution patterns of endophytes have been analyzed to detect fungi that would be likely candidates as biocontrol agents (see Dorworth and Callan this volume). In an attempt to localize a dominant endophyte of *Abies balsamea* (L.). Mill. that might act as a microbial antagonist of the gall midge *Paradiplosis tumifex* Gagné colonization frequencies of dominant endophytes were compared between galled needles and healthy needles. Healthy leaves and young stems (2-3 yr old) of forest weed species, *Acer macrophyllum* and *Alnus rubra,* were surveyed for indigenous latent pathogens that might serve as mycoherbicides (120). Several potential pathogens were identified in each host, but *Melanconium apiocarpum* Link was singled out as a potentially pathogenic endophyte that merited further pathogenicity studies as a control agent for *A. rubra* in the understory of tree plantations (121).

The inherent bioactivity of many phytotoxic metabolites points to endophytes and phytopathogens as sources of useful microbial metabolites (11). Different groups working in large industrial screening programs have examined endophytic communities emphasizing methods for sampling, effects of isolation media and culture technique, maximizing taxonomic diversity of isolates, and evaluation of host specificity (16, 18, 43, 45, 100, 143).

Transportation and Storage of Plant Organs

The logistics of moving fresh plant materials to culturing facilities while maintaining tissues in good condition can be a formidable job when working with plants from remote areas. Precautions must be taken in the shipment of living plant material for culture isolation purposes so the need to prevent desiccation and tissue death is balanced with the requirement for proper ventilation. Aeration maintains respiration, prevents overhumid conditions and suppresses growth of epiphytic fungi and bacteria. At the same time, it is essential that if materials are moved across any regulated political border that all packing criteria established by government agencies (e.g. U.S. Fish and Wildlife, U.S.D.A. Animal and Plant Health Inspection and U.S. Postal Services in the United States) and applications for permits are fulfilled.

There always exists the possibility that the tissues of interest could have been infected by fungal propagules after their collection but before the surface sterilization procedure. Ideal conditions for germination of epiphytic caulo- or phylloplane fungi can be created by maintaining materials in humid environments at ambient temperatures (80). Recently penetrated hyphae or haustoria could survive surface sterilization procedures (138). Epiphytic fungi, e.g. *Alternaria alternata* (Fr.) Keissl. and *Cladosporium cladosporioides* (Fresn.) DeVries may form limited infections through substomatal chambers and thus avoid the effects of surface sterilization (23). Therefore, prolonged transport in sealed plastic bags should be avoided if possible. Avoiding overhumid storage conditions and elevated temperatures is nearly impossible when collecting plants in remote humid tropical forests. When plants are collected under wet tropical conditions, surfaces should be thoroughly air-dried prior to packing and shipping. Bernstein and Carroll (12) recommended transporting conifer foliage in unclosed bags for local transport. Baird (7) maintained spatial position and orientation of bark plugs between the field and laboratory by ordering the plugs in precoded sterile microtiter plates. However, long distance shipping of tropical plants in polyethylene bags has yielded acceptable results (45, 100). Muslin bags were used for long-distance shipping of live Australian fan palm leaves (*Licuala ramsayi* (Muell.) Domin) (108). Sturdy paper envelopes, muslin or Tyvek® soil bags are useful for long-term transport. If plants are stored long periods, especially in frost-free refrigerators, tissue desiccation will occur with a subsequent loss of some internal fungal species. Still, a surprising number of species can be isolated from desiccated woody tissues even after freezer storage of more than a year (Bills & Polishook, unpublished).

Table 1. Selected examples of isolation procedures for endophytes from woody plants.

Tissues/hosts (reference)	dissection	surface sterilization	media	comments
spines and stems of *Ulex* spp. (50)	spines and 2 cm stem sections	1 min 96% ethanol, 3 min 3.25% NaOCl, 0.5 min 96% ethanol	malt extract agar	frequency of colonization increased from new to 2 yr old stems
bark of *Fagus grandifolia*, (37)	2 mm borer	propane torch or none	bark extract, glucose - yeast extract, benomyl malt extract	bark plated in the field
xylem and bark of *Alnus* spp. (53)	1 cm stem sections, bark and xylem cultured separately	35% peracetic acid	malt extract with 10 mg/l cyclosporin A	multivariate analysis revealed bark and xylem and the 3 different *Alnus* spp. can be distinguished on the basis of fungal communities
bark of *Castanea dentata*, *Quercus rubra* (7)	5 mm arch punch	0. 525% NaOCl for 10 minutes	glucose-yeast extract medium	microtiter plates for maintaining sequence of bark samples
mature stems of *Picea abies* (110)	sawing followed by chisel extraction	only extraction tools sterilized	malt agar in slants	frequency of isolates were enumerated along transverse and longitudinal axes of sound and wounded wood
roots of *Picea abies* (68)	1 cm segments	sink washing, ultrasound, serial washing	malt extract agar	3 primary fungal assemblages identified that were correlated with site altitude and edaphic factors
twigs of *Fraxinus excelsior*, *Quercus robur*, *Fagus sylvatica* (64)	2 cm segments, bark and xylem cultured separately	ethanol flaming	malt extract agar	succession of decomposition followed after stem girdling
bark of *Carpinus caroliniana* (16)	1 cm leather punch	0. 525% NaOCl for 3 minutes, flaming in alcohol lamp	malt yeast extract agar with or without benomyl and surfactants, Mycosel agar	imperfect surface sterilization compared with sequential 3-step procedure, use of quartered Petri dishes
seedlings of *Rhizophora mangle* (87)	1.1-2 cm cork borer of hypocotyl and radicle	sterile sea-water rinse, 0.1% HgCl2 in 5% ethanol	mangrove-sea water agar, 4% mangrove ground mangrove seedling in sea water	succession of mycoflora followed in seedlings maintained in nylon mesh bags
leaves of *Licuala ramsayi* (108)	3 mm discs from veins and interveins	1 min 96% ethanol, 10 min 3.25% NaOCl, 0.5 min 96% ethanol	cornmeal dextrose agar, malt extract agar	leaves shipped in muslin bags, higher frequency of endophytes in veins, endophytes in veins of unopened leaves
leaves of *Euterpe oleracea* (109)	3 mm discs from veins and interveins	1 min 75% ethanol, 10 min 3.25% NaOCl, 0.5 min 75% ethanol	cornmeal dextrose agar with 5 mg/l cyclosporin A	leaves processed within 24 hr of collecting

Surface Sterilization

Probably no other step is as critical to obtaining good results as thorough but non-penetrating surface sterilization. The possibility that isolates have been initiated from propagules on the surface must be minimized. The choice of sterilization times, concentrations, and volumes will be dictated by the thickness of sample, the relative permeability of its surface, and the texture of its surface (Table 1). Selection of isolates that have emerged from the cut ends of surface sterilized grass leaves has been practiced to ensure that only internal fungi were selected (79). Such strict criteria usually have not been applied to isolations from woody plants.

Sterilization methods continue to vary widely, but the preferred method is a three-step ethanol, sodium hypochlorite (NaOCl), ethanol treatment (97). Independent tests demonstrated that sequential treatment with ethanol-NaOCl-ethanol effectively killed the thick-walled ascospores of *Sporormiella intermedia* (Auersw.) Ahmed & Cain (96), and conidia of *Heliscus lugdunensis* Sacc. & Therry, *Tricladium splendens* Ingold and *Cylindrocarpon destructans* (Zins) Scholten (56). It should be noted that sodium hypochlorite solutions spontaneously decompose in storage and therefore percentages calculated from labels of household bleach should not be considered accurate. Hydrogen peroxide solutions, commonly used as disinfectants, have not been widely used in endophyte isolation studies. Surface sterilization with 0.05 to 1 % peracetic acid in 30% ethanol may be a more accurate alternative to sodium hypochlorite solutions (M.M. Dreyfuss, personal communication). Ethanol (96%) followed by mercuric chloride (0.1%) and sterile H_2O was used to surface sterilize leaves of *Eucalyptus viminalis* Labill. (13). Leaf surfaces of *Populus tremuloides* Michx. were disinfected by silver nitrate (0.1%) followed by sodium chloride rinses (144).

Ethanol immersion followed by flaming can also yield reasonably good results with larger diameter twigs, limbs and wood fragments. Stems can be dipped in ethanol and ignited with a Bunsen burner or alcohol lamp (64, 119) and whole stem pieces used or the bark and underlying xylem cultured separately. Flaming bark of *Fagus grandifolia* in the field with a propane torch before chip extraction reduced the isolation frequency of moniliaceous hyphomycetes (e.g. *Penicillium* spp., *Fusarium* spp., *Cladosporium* spp.) and algae (37).

Serial washing offers the advantage of eliminating the penetrating and killing effects of sterilizing chemicals. Prewashing with tapwater can help reduce the time needed for surface sterilization. This is especially important if very tiny fragments of tissues are being used. Washing also may be used in conjunction with ultrasonic cleaning and/or detergents. However, with washing alone, in all likelihood, some colonies will arise from propagules embedded in surface irregularities. A serial washing procedure based on the classic soil-washing technique (92) was used to remove phylloplane fungi and epiphytes from pine needles (74). To follow fungal succession in leaves of *Nothofagus truncata* (Col.) Ckn. (114), leaf discs from newly emerged, mature, senescent, and dead leaves were either serially washed or surface-sterilized 1 minute with a solution of 0.01% $HgCl_2$ and 1% "industrial alcohol" and plated onto dextrose-peptone agar. When compared to untreated leaf surfaces, serially washed and surface-sterilized leaf discs both yielded comparable results in terms of frequency of species isolated. Both kinds of surface treatments demonstrated that leaves were colonized by a distinctive group of internal parasites and saprobes while still living on the tree and that many of these species were actively colonizing senescent and dead leaves before being incorporated into the litter. Roots of *Picea abies* (L.) Karst. were cooled in an icebath and cleaned by ultrasonification and serial washing prior to isolation of endophytes (68).

Selective Antibiotics and Nutrients and Culture Manipulation

The phytopathological literature abounds with media formulations designed to recover targeted pathogens from various substrates (132, 135). These kinds of media may be useful if the goal is an autoecological study of a single endophytic species. Frequently however, the primary objectives in studying endophytic communities require media and isolation procedures that recover the broadest range of internal fungi, while discouraging or eliminating interference by external fungi. Considerable attention has been given to the subject of media for isolation of soil fungal communities, but as yet comparatively little has been done in the realm of media development for isolation of endophytic communities. However, this may reflect the fact that those people working in the field are generally satisfied with the standard media described below.

The majority of mycofloristic studies of complex organic substrates are based on methods using so-called "non-selective" media, which are in reality highly selective. In other words, the methods and media create conditions that favor recovery of species with robust and rapid radial growth and high numbers of propagules in the substratum. What is really meant by "non-selective" is those isolation media are chosen that will permit the growth of most species of fungi once they are obtained in a pure culture. The problem faced by one that desires to examine a complete mycoflora is that a set of isolation conditions is needed that will permit equal expression of the entire array of fungal groups present, in so far as they are able to grow at all in artificial culture. Because media that are nutritionally enriched enough to support slow-growing fungi invite inundation by rapidly growing and heavily sporulating fungi, it then becomes necessary to eliminate or restrict the latter by physical or chemical means. Thus, to achieve the isolation of the total culturable fungal community, destructive chemical and physical procedures must be in equilibrium with the need to initiate colonies from vegetative structures inhabiting internal tissues.

Incorporation into agar media of sublethal doses of fungitoxic agents that restrict radial growth of colonies, either singly or in combination, can effectively suppress rapid-growing epiphytic fungi, such as *Trichoderma, Penicillium, Alternaria*, or *Cladosporium* spp. Similarly radial growth rates range widely among different endophytes. Fast-growing endophytes such as *Nodulisporium, Phomopsis, Pestalotiopsis, Botryosphaeria* or *Fusarium* spp. will often overwhelm plates. Antibacterial antibiotics should always be included in any primary isolation medium for fungi. Oxytetracycline, chlorotetracycline, streptomycin sulfate, and novobiocin have been used most frequently for endophyte isolation. Chloramphenicol is an especially easy to use antibacterial agent because it is relatively heat stable and withstands autoclaving (4).

The most effective compounds are those that restrict or distort hyphal extension without causing cell death. One example of such a compound is cyclosporin A (active component of Sandimmune®, Sandoz Pharma, Ltd.) which was originally discovered because of its selective antifungal properties against a variety of filamentous fungi (20, 42). The mechanism of action of cyclosporin A on fungal hyphae is unknown but both its immunosuppressant and antifungal effects appear to be mediated by highly specific cyclosporin-binding proteins, the cyclophilins. Cyclosporin binds to cyclophilin to form a drug-receptor complex that inhibits the Ca_2+ dependent activation of transcription mediated by calcineurin. In filamentous fungal cells, the interaction of cyclosporin and cyclophilin have been shown to modulate lac-1 transcription (76) and to interfere with a-mating factor arrest in yeasts (57). The morphological alterations caused by cyclosporin superficially resemble those of cell wall active antifungals such as nikkomycin and the echinocandins. Even at very low concentrations in agar (1-10 mg/l), fungal hyphae become very distorted, swollen, and exhibit abnormal branching (43). Originally its use was suggested for increasing colony density (and therefore species diversity) on fungal isolation plates during large scale isolation of fungi for metabolite screening because it causes colonies of ascomycetous fungi to become unusually compact and cushion-like and restricts spread and sporulation of mucoraceous fungi (43). The compound has been used successfully to aid fungal isolation from twigs (53), leaves (109), and lichen thalli (103).

Several other fungitoxic compounds have shown potential value for selective isolation of endophytes. The fungicide dichloran (2,6-dichloro-4-nitroaniline, active ingredient of Botran®) can be used effectively at 2 mg/l in agar media (4, 146). Cycloheximide, a protein synthesis inhibitor, has been used to selectively isolate *Ophiostoma ulmi* from diseased vascular tissue of Dutch elm disease infected *Ulmus* spp. (127) and *Leptographium* spp. from conifers and conifer wood (65). Cycloheximide is the selective agent in prepared media (BBL Mycosel agar, Difco Mycobiotic agar) designed to isolate dermatophytes from clinical samples. Preferential isolation of several dematiaceous hyphomycetes and *Beauveria bassiana* was achieved from live bark using Mycosel agar at 0.4 g cycloheximide/l (16). However, when used with small diameter twigs or leaves, the toxic effects depressed isolation frequencies (18). Sodium propionate and bile salts, surfactants long-used for isolation of soil fungi, may be useful as colony restricting agents (16). Cooke's medium (36) with 35 mg/l rose bengal was used to isolate endophytes from *Eucalyptus* leaves (13).

As new antifungal agents with modes of actions that specifically interfere with different sites of cellular function become available, the possibilities for introduction of selective agents into isolation media become infinite. An excellent example of how concentrations of different selective additives were optimized before field testing was presented by Worrall (146) as he tested combinations of benomyl and dichloran to provide a more satisfactory medium for isolation of wood-decay basidiomycetes. Application of selective nutrients in isolation media for endophytes, such as cellulosic materials, unusual sugars or polysaccharides, organic acids, or plant extracts, rarely have been examined in any systematic fashion. Likewise, the effects of media pH on isolation frequencies of endophytes have not been studied. Chi-square analysis of isolation frequencies from live bark demonstrated that isolation media of beech bark extract and potato-sucrose recovered a wider variety of fungal genera than the more selective benomyl-malt extract medium, but a combination of the three media was useful in obtaining a greater variety of taxa (37).

"Divide and conquer" is another tactic used to diminish the assault of fast-growing forms. Espinosa-Garcia and Langenheim (48) remarked that placing five leaves of *Sequoia sempervirens* together in the same plate may have caused frequencies scores of some slower species to be underestimated due to interference from fast-growing species. Segregating fewer pieces into more plates, although consuming more lab supplies, reduces interference from adjacent colonies. Individual segments can be plated in 60 mm Petri dishes (50, 101). Segments placed in compartments of carefully poured quartered Petri dishes is another separation strategy (16, 18). Another effective way to separate colonies is to eliminate or weed out unwanted colonies or remove colonies from which isolates have already be taken. Weeding can be especially effective if done during the first few days of incubation. Colonies can be cut out with a small sharp scalpel or a needle. A high-temperature soldering gun equipped with a fine tip can also be used to eliminate rapidly spreading colonies when they first appear on isolation plates (43, 103).

Mite (*Tyroglyphus* spp., *Tarsonemus* spp.) infestation of cultures is always a threat when working with endophytic fungi because living plant material is being brought into the lab. Also, the prolonged incubation and storage of sterilized plant material covered with mycelium and sporulating structures multiplies the opportunity for mite infestion. Several mite control methods for pure cultures have been devised (123, 135). An especially effective and convenient method for deterring mite infestions is to incorporate minute quantities of insecticide directly into agar media or moist chambers with incubating plant material. For example the chlorinated insecticide, dieldrin, has little effect on fungal growth and can be added directly to agar media or to moist chambers with raw natural substrates. A stock solution of dieldrin (20 mg/ml) in acetone is prepared and then 1 ml can be added to a liter of medium prior to autoclaving to yield a final concentration of 20 mg/l. A drop can added directly to moist chambers (J. Cano and J. Guarro, personal communication). Care must be taken not to contaminate skin, clothes or the laboratory with dieldrin since it is highly toxic. All cultures and materials should be treated as hazardous waste when experiments are finished.

Because of a desire for experimental uniformity, there is tendency to adhere to one constant temperature for isolations. However, when handling large numbers of isolation plates with many colonies, it may be convenient to refrigerate plates or lower the incubator temperatures intermittently while working with other sets of plates, so they are not lost to fast-growing fungi. Most studies are carried out at near room temperatures of 20-25 C, but cooler incubation temperature at 15-20 C may aid to equalize radial growth rates between rapidly growing epiphytes or soil fungi and slower internal forms. If aquatic hyphomycetes are sought, the cooler range is desirable because some of these forms have cooler growth optima (141). Surface-sterilized bark and xylem sections of aquatic roots of *Picea glauca* (Moench) Voss were incubated at 18.5 C to isolate endophytic aquatic hyphomycetes (124). Petrini (98) carried out isolations from leaves of the alpine shrub, *Loiseleuria procumbens*, at 17 C. Ascospores of *Hypoxylon diathrauston* Rehm from high elevation in Switzerland required freezing treatments for germination and optimal growth rates of cultures were at 10-12 C and the maximum was at 18-21 C (90).

Finally, the importance of manual dexterity, constant vigilance and patience should not be underestimated. These virtues, along with good microscopy equipment and minute dissecting tools (viz. insect pins, surgical scalpels), can greatly enhance isolation results by increasing opportunities for isolation of fungi that would otherwise go unnoticed or be consumed by fast-growing species.

Media for Sporulation and Growth

As a rule, media used for routine culture of phytopathogens will be equally applicable to endophytic isolates. Generally prevalent among studies of endophytes are various formulations of malt agar. When handling many unknown species derived from isolation from vegetative growth a useful approach is screen each isolate on several media simultaneously (e.g. cornmeal agar, oatmeal agar, V8 juice agar, etc.) in 60 mm plates. The advantage is that one of the media may permit differentiation of sporulating structures. Also, mixed cultures may be detected because of differential responses of fungi, yeasts or bacteria to selected nutrients. Mycelia on 60 mm plates can be easily scanned for sporulation with a low magnification objective or long-working distance objective from the stage of a microscope. This initial screening can signal the disposition of isolates so that more specialized media for identification or experimentation can be selected. Development of sporulating structures in some endophytes is extremely slow, if sporulation will occur at all. Therefore, maintaining the original isolation plates and surveying them, especially the host tissues, for as long as possible is recommended.

Illumination is necessary to stimulate sporulation in many fungi and often the most effective wavelengths are in the near ultraviolet range. On the basis of Leach's work (77), the International Mycological Institute's identification service recommended illumination of incoming unknown cultures with a combination of near ultraviolet fluorescent lamps and cool white fluorescent lamps on a 12 hr photoperiod (70). Petrini (97) recommended extended incubation at low temperature (4-8 C) either with or without light.

As endophytes continue to be explored for their potential biotechnological applications, traditional media development and scale-up practices to maximize product yield or titer are being applied to these fungi. The degradative capacities and nutritional requirements for in vitro growth, development, sporulation of endophytes can largely be predicted based on previous knowledge of phytopathogenic ascomycetes. Already some optimization experiments of growth and fermentation conditions for specific processes have been carried out using traditional shake flask and fermenter technology for a few endophytic fungi. In the future however, certain endophytes may pose some challenges as the need arises to perform inoculum development and classic genetical and strain improvement techniques with sterile and slow-growing taxa with peculiar growth kinetics.

An understanding of cultural nutritional behavior of a fungus is necessary to understanding its ecological function. The ability of a variety of endophytes to hydrolyze various polymeric components of plant cells has been tested in agar cultures (28, 121, 122), therefore it seems obvious that these organisms should be examined for enzymes useful in commercial food processing. For example, the chestnut-blight pathogen, *Cryphonectria parasitica*, produces an acid protease that is used as a substitute for natural rennet in certain types of cheese-processing (14). Sapwood-colonizing *Hypoxylon* and *Xylaria* spp. have been screened for their ability to excrete cellulolytic enzymes by assaying extracellular endo-β-glucanase, exo-β-glucanase and β-glucosidase activity from shake cultures (142). Cultures were grown in a liquid cellulose-peptone medium. Although cultures were incubated at 28 C, optimum enzyme activities were observed between 37-50 C. Some of the isolates were able to hydrolyze crystalline cellulose more rapidly than *Trichoderma reesei*. One isolate of *Hypoxylon stigium* possessed an extremely potent b-glucosidase, completely hydrolyzing the cellulose of the liquid medium in 3 days. Rapid colonization of bark of *Populus tremuloides* by *Hypoxylon mammatum* (Wahlenberg) J. H. Miller may be related to the stimulatory effect of proline on the growth rate of the fungus (62). Depending on the isolate, proline, especially in combination with other amino acids such as glutamine in semi-synthetic media, increased rates of hyphal extension up to 2-4 times compared to other amino acids. Proline accumulated in tissues of the water-stressed host. Therefore, it was suggested that drought-related canker elongation may be related to proline content of the bark.

Bioassay-guided fractionation of fermentation extracts has directed chemical purification of bioactive metabolites from endophytes. Three new cytochalasins were discovered in *Hypoxylon fragiforme* because one of them, L-696,474 (18-dehydroxy cytochalasin H), was a potent competitive inhibitor of HIV-1 viral protease (41, 89). Cytochalasin production could be achieved on a variety of solid and liquid media, however best yields of L-696,474 were obtained in liquid shake cultures with a complex medium based on sucrose, glycerol, citrate, ardamine (a yeast extract preparation), soybean meal, and tomato paste. Echinocandin analogues, potent lipopeptide inhibitors of fungal β-1,3 glucan synthase, have been isolated from *Cryptosporiopsis* spp. (88, 133). Chemical isolation, identification and antifungal testing of one echinocandin analogue, L-671,329 (pneumocandin A0), were carried out from small shake flasks containing a complex medium of maltose, maltodextrins, peptone, cottonseed flour, molasses and mineral salts (88). The tremorgenic neurotoxins, paspalitrem A and C, previously only known from sclerotia of *Claviceps paspali* F. Stevens & Hall, have been produced by an endophytic *Phomopsis* by solid fermentation on millet seeds (15). An unusual, cytotoxic and fungitoxic aromatic bis-ketal, preussomerin D, has been isolated from *Hormonema dematioides* Lagerberg & Melin grown on a solid rice-based medium (105). A group of novel cytochalasins was produced by *Hypoxylon terricola* J. H. Miller after stationary growth for 8 weeks in 500 ml of 2% malt extract in 1 liter Thompson bottles (46).

A few endophytes have now been grown in large scale bioreactors for production of antibiotics. A novel lactone with potential antiherpes and antimicrobial properties was discovered in an endophytic *Microsphaeropsis* strain (134). To obtain enough purified compound for testing, it was necessary to develop fermentation protocols for bioreactors of up to 5200 liters. A new cyclodepsipeptide antihelminthic (115) was produced by a sterile fungus isolated from *Camellia japonica* L. A three-stage scale-up from 100 ml shake flasks to 50 liter jar fermenters was employed to obtain sufficient material for chemical isolation and characterization.

Preservation Methods

The methods of storage of endophytic isolates will depend upon the intended use of the cultures, facilities, and space. If repeated use over an extended time is anticipated, for example for physiological, pathogenicity, or fermentation studies, then well stabilized stocks are essential. Any extensive study of endophytic fungal communities will result in an accumulation of a large number isolates of many species with varying capacities for growth and survival in culture. Because endophytes are isolated on agar media, most grow, often readily on agar, and can be maintained on slants at least temporarily.

An immediate problem arises in how to maintain viability of this large quantity of isolates for future studies or as voucher cultures supporting published results. Another problem stems from the variability in fungal forms that can be isolated and their ability to sporulate or produce other reproductive structures that will survive in preservation. Lyophilization remains the most efficient long-term storage method for heavily sporulating fungi. However many isolates will be sterile or otherwise will produce spores in insufficient quantities to survive the lyophilization process. One of the most frequently encountered groups, the *Xylaria* and *Hypoxylon* anamorphs that sporulate in culture, will not lyophilize. Freezing liquid mycelial cultures in 10-20% glycerol at -80 to -100 C or in vapor of liquid nitrogen is rapid, convenient and generally provides extended viability with little culture deterioration, but availability of freezer space or even a freezer may be problematic for many laboratories. Lyophilization and ultracold freezing as preservation methods for filamentous fungi have been extensively reviewed (123, 135). An indication of whether a strain may be amenable to lyophilization can be obtained by first examining the most recent edition of the American Type Culture Collection Catalouge of Filamentous Fungi (72). If the strain is shipped as a lyophilized ampule, then in all likelihood, it can be lyophilized if cultures are in good sporulating condition. If strains are shipped on agar slants, then they probably were stored in liquid nitrogen vapor prior to transfer to agar.

Storage of mycelium on agar plugs in distilled water may be a practical, low-cost preservation alternative for many laboratories. A complete and effective system utilizing small agar plugs of fungal mycelia stored in small plastic cryovials of sterile water and stored at room temperature was recently described (71). The system offered several advantages in that no special, costly apparati were needed (e.g. liquid nitrogen freezers, lyophilizers), a broad diversity of fungi could be preserved, and contamination was minimal due to the inertness of water as a storage medium. A high percentage of viability was maintained after 2-5 years' storage, and the cultures remained in a physiologically stable condition suitable for fermentation studies. This method provided effective preservation for *Mycoleptodiscus atromaculans* Bills & Polishook, a fungus that does not survive well in agar culture (17).

Storage in sterile soil contained in screw cap bottles or tubes, is another low-tech, low cost method. Mycelia and/or spores can be scraped from agar cultures, mixed thoroughly with moistened, double-sterilized soil, incubated a few days to colonize soil particles, and then can be refrigerated. We have used this method in our laboratory for a variety of endophytic fungi, including many nonsporulating strains, with satisfactory results. However, some fungi, such as *Fusarium* spp., that are inherently genetically unstable may degenerate in soil storage (145).

Voucher cultures of unusual taxa, new species, or isolates used for biotechnological applications should be deposited in permanently maintained culture collections. Specimens collected from the field, dried culture mats and/or permanently mounted lactophenol slides conserved in recognized herbaria may also serve as vouchers in place of cultures and are required if new taxa are described. Such collections are valuable references for one's future work or for other researchers who will want to compare their isolates with those from previous studies.

Identification

This section is intended primarily for those entering into the murky waters of endophyte taxonomy for the first time. For the novice and the experienced, the most difficult and time-consuming, but vital part, of a fungal community analysis will be identification of the isolates. Experienced fungal biologists are all too familiar with difficulties of identification and realize many of the classic problems of fungal systematics are exaggerated when faced with the prospect of naming dozens of species based solely on features expressed (or not) in culture. Because of the diffuse nature of fungal taxonomic literature, identification of dozens of different taxa from a single host will require repeated entry of the taxonomic literature at several disjunct points (49, 113).

The good news is that when one is working with isolates from a single host plant over a short duration of time "morphological species", regardless of their name, can be defined relatively easily and quickly sorted into preliminary taxa. In our laboratory we have taken a straightforward approach described for study of soil fungal communities (35). Isolates with similar provenances are point inoculated at the apex of agar slants. Agar slants (e.g. malt extract, yeast malt extract) should be uniform in age and composition, inoculated at approximately the same time, and incubated together. After a few weeks incubation, assignment of morphological species can be based on colony surface textures, hyphal pigments, exudates, margin shapes, growth rates, and sporulating structures. With this simplistic and mechanical approach, comparisons in floristic composition and species frequency can still be made among small groups of related samples without needing to name the isolate. But applications of this approach are limited because the kinds of taxa involved can not be compared over time or communicated with other researchers.

Because fungal names are tied to sporulating structures and because the majority of plant-inhabiting fungi have been named based on structures produced on the host, examination of the host for fungi is a logical starting point. If an extended investigation of a single host plant is anticipated, then before starting isolations, it may be worthwhile to gather data on what fungi should be expected. Reference to host indices (47, 49, 58, 136, 112) and identification manuals directed toward specialized plant groups for certain geographic regions (22, 47, 58, 59), where available, can indicate what some of the common fungi will be beforehand. However, the predictive value of these works is generally limited to fungi of major forest species of the northern hemisphere and of important crop plants.

Collections of recently dead twigs, stems, and leaves and litter should be examined for sporulating structures. Senescent or dead plant parts sometimes can be rehydrated and incubated in moist chambers to resume active sporulation. Moist chamber incubation or slow drying of living host tissues to induce development of lesions and emergence of sporulating structures can be carried out in parallel with isolations on nutrient agar to induce sporulation on host tissue (34). Incubation of woody stems in plastic bags impedes desiccation and can permit mycelial outgrowth and identification or isolation of mycelium from colonized regions of decorticated stems or external fruiting on corticated stems (32). The cut ends of limbs and branches can be sealed to impede desiccation and then incubated to permit development of fruiting structures on the wood surface. Anderson and French (1) observed the development of *Hypoxylon mammatum* on field-collected stems of *Populus tremuloides*. *Hypoxylon* cankers developed on stem sections that were coated with paraffin and incubated in sterile wet sand in greenhouses or incubators. Chapela and Boddy (33) and Chapela (31) used a similar approach to determine the effect of desiccation on colonization patterns by latent fungi in cut limbs. Distilled water agar slabs sealed in moistened cotton wool and plastic film or cotton wool and punctured plastic film alone were applied to cut stem ends to impede drying during incubation in environmental cabinets or greenhouses. Washed leaf discs of *Populus tremuloides* were incubated in moist chambers. The fungi developing on leaf discs were compared with those obtained from surface-sterilized discs plated on malt agar (144). Chemical induction of senescence with paraquat may also provoke sporulation on the host (29; Sinclair & Cerkauskas, this volume).

Girdling (ring-barking) living stems (64, 106) may release latent endophytes in bark and sapwood and encourage their development in situ. Sudden death of oaks caused by oak wilt or girdling, provided Barnett (9) the opportunity to characterize the life cycle of *Hypoxylon punctulatum* (Berk. & Ravenel) Cooke, a primary colonizer of the recently killed oak stems. He observed the intimate association of the *Basidiobotrys* (now classified in the form genus *Xylocladium*) state with emerging stromata and proved their connection through cultural studies. The conidial state developed on internal pillars within the stromata and could be isolated from wood beneath the bark. Emergent stromata developed from deep (up to 6 mm) in the bark. After following the ontogeny of the stromata, a mechanism was proposed of how gelatinous tissues inside the stromata expanded to eject the outer bark so as to expose the fertile perithecial layer.

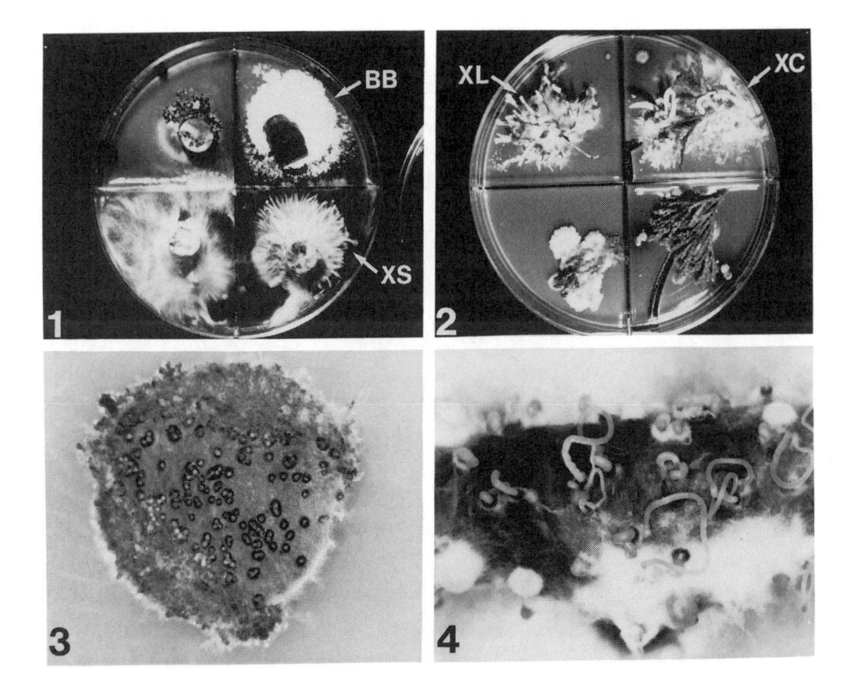

Figs. 1-4. Sporulation of endophytic fungi on different types of healthy plant tissue in agar culture. 1. Endophytes emerging from bark discs of *Carpinus caroliniana* after 2 weeks on malt yeast extract agar. BB = *Beauveria bassiana*, XS = unidentified *Xylaria* anamorph. 2. Endophytes emerging from leaf sections of *Chamaecyparis thyoides* after 3 weeks on Mycosel agar. XL = anamorph of *Xylaria longipes* Nitschke, XC = anamorph of *Xylaria cubensis* (Mont.) Fr. 3. Leaf disc (5 mm) of *Ilex opaca* with emergent conidiomata of *Phyllosticta concentrica*. After 2 weeks on malt yeast agar with 4 mg/l cyclosporin A. 4. Conidiomata of *Cytospora salicis* (Corda) Rabenh. developing from twig section of *Salix alba* L.

During prolonged incubation of twigs and leaf sections on the primary isolation plates, some taxa will preferentially sporulate on host tissue (Figs. 1-4). Extended incubation of twig and needle sections permitted development of apothecia of *Chloroscypha* spp. in culture. (95). Unknown sporulating structures observed in cultures, either on plated host tissue or produced on agar, can be matched with those found on host tissues in nature. *Rhabdocline parkeri* was first observed as sporodochia protruding through stromata and apothecia forming on surface-sterilized needles of *Pseudotsuga menziesii* in culture. Later, sporodochia were compared with structures collected on galled needles infested by the gall midge, *Cortarinia pseudotsugae* Condrashoff and with apothecia forming naturally on recently abscised needles (118). Sridhar and Bärlocher (125) observed formation of the teleomorph of *Heliscus lugdunensis* Sacc. & Therry after 40 days of incubation after being subcultured from bark of *Picea glauca* (Moech.) Voss. Inoculating cultures back on host tissue or other plant material may induce sporulation, allow for more natural development of sporulation, or permit observation of teleomorphs or synanamorphs not observed on agar (91, 97, 137).

Many species of endophytes will not sporulate in agar culture, while with cultures of many others, the ability to sporulate deteriorates rapidly with successive transfers. In other cases, the time needed for sporulation may be weeks or months. Therefore isolates need to be identified and characterized as soon as possible. Carroll (26) asserted that a century's "accumulated lore on manipulation of growth media and laboratory environment to induce sporulation" in endophytic isolates should be applied to sterile isolates. But which folkloric remedy should be applied to which isolate? This approach may be valuble if one knows beforehand something of the biology of the isolate, and if there is sufficient reason to proceed with the tedious screening of each single isolate though countless media and incubation conditions. The ideal situation would be similar to that described by Barnett (8) where a semi-synthetic glucose-phenylalanine medium was developed that was suitable not only for direct isolation of the the oak-wilt fungus from living wood but promoted its rapid sporulation and identification. Even when there is potential economic incentive to experiment with growth conditions, such as an isolate producing an interesting antibiotic, many sterile isolates are doomed to remain sterile.

The technological capacity to determine the approximate phylogenetic position of any sterile isolate among extant fungi is now at hand. Construction of partial phylogenies of ascomycetous fungi has been achieved by sequence analysis of enzymatically amplified DNA copies of ribosomal RNA genes from an array of representative taxa (10, 21). Through alignment and cladistic analysis with homologous nucleotide sequences of known fungi, phylogenetic relationships can be inferred and the unknown sterile strain could be assigned to a higher taxonomic category (class, order or family), even without positive assignment of names. In this way the phylogenetic diversity of endophytes could be assessed independent of taxonomy. With knowledge of the approximate phylogeny of unknown isolates, conditions for manipulation of growth and sporulation could then be predicted.

Coelomycetes will be among the most frequent and the most problematic taxa encountered. The identification of endophytic coelomycetes is especially frustrating because scarcely any cultural data is available for many of the predominant taxa. Pleomorphy among pycnidial, sporodochial, hyphomycetous, and yeast-like states is common. If possible, identification should be carried out from structures developing on host material because descriptions of the majority of these fungi are available only from sporulating structures collected in nature. The limitations of applying morphological descriptions of coelomycete taxa made exclusively from cultural studies has been emphasized previously (86). Except for a few genera, e.g. *Colletotrichum* and *Phoma*, virtually all identification schemes are based on morphological features developed on the host.

Another major group that is consistently encountered in most plants are species in the Xylariales (97). Humid tropical forests appear to be a center of radiation for these fungi where they are especially common and diverse in living plants (107), dead wood and plant litter. When isolated from vegetative mycelium, generally only anamorphic states form in culture. The form genera are relatively easy to distinguish, but species identification in many groups is nearly impossible. However, through careful studies of cultures derived from named teleomorphs many of the more common temperate species can now be identified from culture (93).

Closely related species and strains isolated from same or different hosts have been analyzed in detail to determine if species or strains isolated on different hosts evolved independently. Evidence based on somatic compatibility testing, comparisons of physiological characteristics and enzymes indicate that strains of the same endophyte species may have evolved host-specific races (78, 121, 122) or may be composed of genetically heterogeneous populations (107). Even single spore isolates of a species from the same host may exhibit broad physiological and morphological variation, as evidenced by variation among isolates of *Hypoxylon mammatum* in their specific radial growth rates on different amino acids (62). Morphometric analysis of spore volumes with an automated particle sizer (30, 131) is an extremely innovative technique that has been applied to the age-old problems of detecting minute morphological differences among individuals of complex taxa. No doubt this technique will find many uses as a rapid and statistically powerful technique for comparing distributions of spore dimensions in other groups of fungi.

Endophytes and Exploration of Fungal Biodiversity

Dreyfuss (44) speculated that endophytic fungi represent one of the largest reservoirs of fungal species. This proposal seems plausible based along two lines of reasoning. Isolation studies indicate that each species of vascular plant may harbor at least 2-4 endophyte species unique to that plant species. Second, if one scans fungal host indices (49, 61, 126) or compilations of fungal genera (136), huge numbers of taxa, especially genera of coelomycetes that are more than likely endophytes, e.g. *Coniothyrium, Cylindrosporium, Stagonospora, Hendersonia, Phoma, Phomopsis, Phyllosticta, Septoria*, etc., occur on nearly all vascular plants observed. Multiply these numbers of fungal taxa by estimates of the total number of vascular plants and the numerical consequences become obvious (66). If only a fraction of these taxa are real and truly host-specific, as new evidence indicates (99), then the estimates of total fungal species involved are of astonishing magnitude.

Until now most studies of endophytes of woody plants have focused on northern temperate forest plants, especially major forest species and ericaceous shrubs. In temperate zones, intensive studies on individual plant species can easily double or triple the number of fungal species associated with that plant when compared to information previously compiled in host indices. Only a handful of studies have documented endophytes in tropical plants (45, 100, 108, 109) and only one tropical forest plant, açai palm (*Euterpe oleracea* Mart.), in the Brazilian Amazon, has been thoroughly investigated (107). The floristic information regarding species diversity gained from these studies can be tremendous relative to the short amount of time and limited plant material necessary to carry them out. Sampling methods are readily standardized to provide comparative data. Considering the pressure for rapid assessment of microbial (fungal) diversity in tropical biomes (67, 130), surveys of endophytes in endemic plants should be given high priority as a method to complement traditional field collecting. Furthermore, because the organisms have been isolated into axenic culture, they have the potential to be maintained permanently for future taxonomic, physiological, or chemical study.

Lack of adequate laboratory facilities in tropical regions will continue to force much of future research on tropical endophytes to be carried out in the temperate industrialized countries. The laboratory facilities and equipment required to carry out such fundamental work are comparatively simple. A standard microbiological laboratory, including a clean room with laminar flow hood, incubators, and equipment for media preparation and sterilization, and microscopy equipment, would be the minimal requisite. Unfortunately, establishment and maintenance of such facilities in many remote areas would be an unsurmountable task, unless they are shared within institutions already engaged in phytopathological, food, or medical microbiological research.

Potentially Unique Endophytes

The overall floristic patterns of endophytic fungal communities of temperate plants are now well-established and discoveries of significant variations in the future seem unlikely. However, detailed exploration of highly modified vascular plants, unusual plant organs or plants with specialized nutritional modes surely will continue to yield interesting new taxa and observations of known taxa in hitherto unknown associations. Some saprobic achlorophyllous seed plants, such a *Monotropa* and achlorophyllous orchids, have highly modified or reduced vegetative organs and unique basidiomycetous fungal symbionts associated with their roots and seeds, but as of yet, the aerial portions of achlorophyllous plants have not been examined for internal fungi. Virtually nothing is known of the mycota of achlorophyllous parasitic plants, such as *Conophilus* and *Epifagus* (Orobanchaceae). Surveys of endophytes in these plants, particularly the haustorial organs, may yield yet unexpected fungal associations.

Adaptations of endophytic communities in plants that grow in extreme habitats, other than arctic or alpine plants, are generally unknown. Desert plants have largely been ignored as sources of endophytes perhaps because data from humid or seasonally humid biomes indicates endophyte infection declines rapidly with decreasing relative humidity and rainfall. In our laboratory, we surveyed leaves of *Yucca* spp. (series *elata*) from New Mexico and Texas for endophytes (Bills & Polishook, unpublished). We used standard isolation procedures (with a 7 min NaOCL rinse) to recover isolates from 24 1-cm leaf segments from each of 50 different plants (1200 segments total). *Alternaria alternata* (Fr.) Keissl. was the most frequently isolated organism (77% of the segments). A variety of sterile strains, coelomycetes, and *Aureobasidium pullulans* (de Bary) G. Arnaud were also among the dominant organisms. Some of taxa that we could definitively identify from culture, albeit occurring at very low frequencies, were *Comoclathris permuda* (Cooke) E. Müller, *Fusarium moniliforme* J. Sheld., *Kellermania yuccigena* Ellis & Everh., and *Periconia maculans* (Cooke) Crane & Schoknecht. Likewise, we have recovered endophytes from living stems or leaves of other perennial desert plant genera such as *Agave, Atriplex, Larrea, Artemesia, Prosopis and Ephedra.* Similar to our observations with leaves of *Yucca* spp. in North American deserts, *Fusarium moniliforme* was commonly associated with roots of the parasitic plant, *Cynomorium coccineum* L., in deserts of Iraq (85). Previous studies of the mycota of desert halophilic plants (84) along with this preliminary data indicates desert plants harbor endophytic infections, but with infections dominated by common dematiaceous epiphytes and with an overall reduced frequency and diversity when compared to plants in humid environments. The interaction of endophytes with plants with highly modified anatomy and physiological adaptations, such as succulent or xeromorphic leaves or salt glands, deserve further study.

Endophytes rarely have been surveyed in intertidal or marine systems. Two species of cool temperate halophytes, *Salicornia perennis* L. and *Suadea fruticosa* Forskål have been examined for tissue-specific endophytes (51, 101). In our own laboratory, we have observed a rich endophyte mycota in leaves and stems of the halophilic shrub *Baccharis halimifolia* L. (Bills & Polishook, unpublished). It is quite likely that many of the ascomycetes characteristic of the submerged portions of *Spartina alterniflora* (60) undergo a latent infection prior to fruiting. Future studies on distribution of endophytes through aerial, intertidal, and submerged tissues of the different species of mangroves (*Avicennia* spp., *Rhizophora* spp., etc.) should present an ideal model for examining fungal infection patterns along a strong anatomical-environmental gradient. Kohlmeyer and Kohlmeyer (75), describing the distribution of fungi on mangroves, stated, "Our observations suggest that undamaged, bark-covered tissues are not invaded by marine fungi." A few of the fungi associated with *Rhizophora mangle* L. seedlings may qualify as endophytes, as defined above. For example, *Cytospora rhizophorae* Kohlmeyer and *Keissleriella blepharospora* Kohlmeyer & Kohlmeyer could be isolated regularly from discs cut from healthy surface-sterilized seedlings and ascocarps of *K. blepharospora* fruited on senescing seedlings incubated in damp chambers (87). Other studies have enumerated frequencies of fruiting structures of various distinctive taxa of marine fungi on intertidal prop roots of mangrove species (69). These quantitative floristic studies show that mangrove-inhabiting fungi exhibit a complex community structure and definite successional patterns. Based on our present knowledge of internal fungi of terrestrial plants and the recent revelations that aquatic hyphomycetes reside as latent infections in roots of submerged fresh water streams, the possibility that parts of the life cycles of marine ascomycetes may persist endophytically in mangroves needs to be examined by direct isolation studies from internal tissues.

Acknowledgments

Appreciation is expressed for the helpful comments and suggestions provided by M. M. Dreyfuss, L. R. Koupal, K. F. Rodrigues and J. K. Stone. Many of the ideas expressed in this review are based on collective experiences shared with my Merck coworkers J. D. Polishook, F. Peláez and L. J. Girimonte.

Literature Cited

1. Anderson, D. L. and French, D. W. 1972. Isolation of *Hypoxylon mammatum* from aspen stem sections. Can. J. Bot 50: 1971-1972.
2. Ando, K. 1992. A study of terrestrial aquatic hyphomycetes. Trans. Mycol. Soc. Jpn. 33: 415-425.

3. Ando, K. and Tubaki, K. 1984. Some undescribed hyphomycetes in the rain drops from intact leaf-surface. Trans. Mycol. Soc. Jpn. 25: 21-37.
4. Anonymous. 1990. Food Mycology. A guide to the use of Oxoid culture media. Unipath Limited, Basingstoke.
5. Aoki, T., Tokumasu, S. and Tubaki, K. 1990. Fungal succession on momi fir needles. Trans. Mycol. Soc. Jpn. 31: 355-374.
6. Bacon, C. W. 1990. Isolation, Culture, and Maintenance of Endophytic Fungi of Grasses. Pages 259-282 in: Isolation of Biotechnological Organisms from Nature. D. P. Labeda, ed. McGraw-Hill Publishing Co., New York .
7. Baird, R. E. 1991. Mycobiota of bark associated with seven strains of *Cryphonectria parasitica* on two hardwood tree species. Mycotaxon 40: 23-33.
8. Barnett, H. L. 1953. Isolation and identification of the oak wilt fungus. West Virginia Agricultural Experiment Station Bulletin 359T: 1-15.
9. Barnett, H. L. 1957. *Hypoxylon punctulatum* and its conidial stage on dead oak trees and in culture. Mycologia 49: 588-595.
10. Berbee, M. L. and Taylor, J. W. 1992. 18S ribosomal RNA gene sequences characters place the the human pathogen *Sporothrix schenckii* in the genus *Ophiostoma.* Exper. Mycol. 16: 87-91.
11. Bèrdy, J. 1989. The discovery of new bioactive microbial metabolites: screening and identification. Pages 3-25 in: Bioactive Metabolites from Microorganisms. M. E. Bushell and U. Gräfe, eds. Elsevier Science Publishers, Amsterdam .
12. Bernstein, M. E. and Carroll, G. C. 1977. Internal fungi in old-growth Douglas fir foliage. Can. J. Bot. 55: 644-653.
13. Bertoni, M. D. and Cabral, D. 1988. Phyllosphere of *Eucalyptus viminalis* II: Distribution of endophytes. Nova Hedwigia 46: 491-502.
14. Bigelis, R. 1991. Food enzymes. Pages 361-415 in: Biotechnology of Filamentous Fungi, Technology and Products. D. B. Finkelstein and C. Ball, eds. Butterwork-Heinemann, Boston .
15. Bills, G. F., Giacobbe, R. A., Lee, S. H., Peláez, F. and Tkacz, J. S. 1992. Tremorgenic mycotoxins, paspalitrem A and C, from a tropical *Phomopsis*. Mycol. Res. 96: 977-983.
16. Bills, G. F. and Polishook, J. D. 1991. Microfungi from *Carpinus caroliniana.* Can. J. Bot. 69: 1477-148.
17. Bills, G. F. and Polishook, J. D. 1991. A new species of *Mycoleptodiscus* from living foliage of *Chamaecyparis thyoides*. Mycotaxon 43: 453-460.
18. Bills, G. F. and Polishook, J. D. 1992. Recovery of endophytic fungi from *Chamaecyparis thyoides*. Sydowia 44: 1-12.
19. Boddy, L. and Griffith, G. S. 1989. Role of endophytes and latent invasion in the development of decay communities in sapwood of angiospermous trees. Sydowia 41: 41-73.

20. Borel, J. F. 1982. The history of cyclosporin A and its significance. Pages 5-17 in: Cyclosporin A, Proceedings of an International Conference on Cyclosporin A. D. J. G. White, ed. Elsevier Biomedical Press, Amsterdam .
21. Bruns, T. D., White, T. J. and Taylor, J. W. 1991. Fungal molecular systematics. Annu. Rev. Ecol. Syst. 22: 525-564.
22. Butin, H. and Peredo, H. L. 1986. Hongos parásitos en coniferas de América del Sur. Bibl. Mycol 101: 1-100.
23. Cabral, D., Stone, J. K. and Carroll, G. C. 1993. The internal mycoflora of *Juncus* spp.: microscopic and cultural observations of infection patterns. Mycol. Res. 97: 367-376.
24. Carroll, F. E., Müller, E. and Sutton, B. C. 1977. Preliminary studies on the incidence of needle endophytes in some European conifers. Sydowia 29: 87-103.
25. Carroll, G. C. 1987. Fungi isolated from gypsy moth egg-masses. Mycotaxon 29: 299-305.
26. Carroll, G. C. 1991. Fungal associates of woody plants as insect antagonists in leaves and stems. Pages 253-271 in: Microbial Mediation of Plant-Herbivore Interactions. P. Barbosa, V. A. Krischik and C. G. Jones, eds. John Wiley & Sons, Inc., New York .
27. Carroll, G. C. and Carroll, F. E. 1978. Studies on the incidence of coniferous needle endophytes in the Pacific Northwest. Can. J. Bot 56: 3034-3043.
28. Carroll, G. C. and Petrini, O. 1983. Patterns of substrate utilization by fungal endophytes from coniferous foliage. Mycologia 75: 53-63.
29. Cerkausksas, R. F. and Sinclair, J. B. 1980. Use of paraquat to aid detection of fungi in soybean tissue. Phytopathology 70: 1036-1038.
30. Chapela, I. 1991. Spore size revisited: analysis of spore populations using an automated particle sizer. Sydowia 43: 1-14.
31. Chapela, I. H. 1989. Fungi in healthy stems and branches of American beech and aspen: a comparative study. New Phytol. 113: 65-75.
32. Chapela, I. H. and Boddy, L. 1988. Fungal colonization of attached beech branches. I. Early stages of development of fungal communities. New Phytol. 110: 39-45.
33. Chapela, I. H. and Boddy, L. 1988. Fungal colonization of attached beech branches. II. Spatial and temporal organization of communities arising from latent invaders in bark and functional sapwood under different moisture regimes. New Phytol. 110: 47-57.
34. Christensen, C. M. 1940. Studies on the biology of *Valsa sordida* and *Cytospora chrysosperma*. Phytopathology 30: 459-475.
35. Christensen, M. 1969. Soil microfungi of dry to mesic conifer-hardwood forests in northern Wisconsin. Ecology 50: 9-27.
36. Cooke, W. B. 1954. The use of antibiotics in media for the isolation of fungi from polluted water. Antibiot. Chemother. 4: 657-662.

37. Cotter, H. V. T. and Blanchard, R. O. 1982. The fungal flora of bark of *Fagus grandiflora*. Mycologia 74: 836-843.
38. Currah, R. S. 1991. Taxonomic and developmental aspects of the fungal endophytes of terrestrial orchid mycorrhizae. Lindleyana 6: 211-213.
39. Dalp,, Y., Litten, W. and Sigler, L. 1989. *Scytalidium vaccinii* sp. nov., an ericoid endophyte of *Vaccinium angustifolium* roots. Mycotaxon 35: 317-377.
40. Dastur, J. F. 1916. Spraying for ripe rot of the plantain fruit. Agric. J. India 11: 142.
41. Dombrowski, A. W., Bills, G. F., Sabinis, G., Koupal, L. R., Meyer, R., Ondeyka, J. G., Giacobbe, R. A., Monaghan, R. L. and Lingham, R. B. 1992. L-696,474, a novel cytochalasin as an inhibitor of HIV-1 protease I. The producing organism and its fermentation. J. Antibiot. 45: 671-678.
42. Dreyfuss, M., Harri, E., Hofmann, H., Kobel, H. H., Pache, W. and Tscerter, H. 1976. Cyclosporin A and C, new metabolites from *Trichoderma polysporum*. Eur. J. Appl. Microbiol. 3: 125-133.
43. Dreyfuss, M. M. 1986. Neue Erkenntnisse aus einem pharmakologischen Pilz-screening. Sydowia 39: 22-36.
44. Dreyfuss, M. M. 1989. Microbial diversity. Microbial Metabolites as sources for new drugs, Princeton Drug Research Symposia, Princeton.
45. Dreyfuss, M. M. and Petrini, O. 1984. Further investigations on the occurrence and distribution of endophytic fungi in tropical plants. Bot. Helv. 94: 33-40.
46. Edwards, R. L., Maitland, D. J. and Whalley, A. J. S. 1989. Metabolites of the higher fungi. Part 24. Cytochalasin N, O, P, Q, and R. New cytochalasins from the fungus *Hypoxylon terricola* Mill. J. Chem. Soc. Perkin Trans. I 1989: 57-65.
47. Ellis, M. B. and Ellis, J. P. 1985. Microfungi on Land Plants, An Identification Handbook. Macmillan Publishing Company, New York. 818 pp.
48. Espinosa-Garcia, F. J. and Langenheim, J. H. 1991. The endophytic fungal community in leaves of a coastal redwood population - diversity and spatial patterns. New Phytol. 116: 89-97.
49. Farr, D. F., Bills, G. F., Chamuris, G. P. and Rossman, A. Y. 1989. Fungi on Plant and Plant Products in the United States. American Phytopathological Society, St. Paul, MN. 1252 pp.
50. Fisher, P. J., Anson, A. E. and Petrini, O. 1986. Fungal endophytes in *Ulex europaeus* and *Ulex galli*. Trans. Br. Mycol. Soc. 86: 153-193.
51. Fisher, P. J. and Petrini, O. 1987. Location of fungal endophytes in tissues of *Suaeda fruticosa*: A preliminary study. Trans. Br. Mycol. Soc. 89: 246-249.
52. Fisher, P. J. and Petrini, O. 1989. Two aquatic hyphomycetes as endophytes in *Alnus glutinosa* roots. Mycol. Res. 92: 367-368.

53. Fisher, P. J. and Petrini, O. 1990. A comparative study of fungal endophytes in xylem and bark of *Alnus* species in England and Switzerland. Mycol. Res. 94: 313-319.
54. Fisher, P. J. and Petrini, O. 1992. Fungal saprobes and pathogens as endophytes of rice (*Oryza sativa* L.). New Phytol. 137-143.
55. Fisher, P. J., Petrini, O. and Petrini, L. E. 1991. Endophytic ascomycetes and deuteromycetes in roots of *Pinus sylvestris*. Nova Hedwigia 52: 11-15.
56. Fisher, P. J., Petrini, O. and Webster, J. 1991. Aquatic hyphomycetes and other fungi in living aquatic and terrestrial roots of *Alnus glutinosa*. Mycol. Res. 95: 543-547.
57. Foor, F., Parent, S. A., Morin, N., Dahl, A. M., Ramadan, N., Chrebet, G., Bostian, K. A. and Nielsen, J. B. 1992. Calcineurin mediates inhibition by FK506 and cyclosporin of recovery of a-factor arrest in yeast. Nature 360: 682-684.
58. Funk, A. 1981. Parasitic Microfungi of Western Trees. Canadian Forestry Service, Pacific Forest Research Centre, Victoria. 190 pp.
59. Funk, A. 1985. Foliar Fungi of Western Trees. Canadian Forestry Service, Pacific Forest Research Centre, Victoria. 159 pp.
60. Gessner, R. V. 1977. Seasonal occurrence and distribution of fungi associated with *Spartina alterniflora* from a Rhode Island estuary. Mycologia 69: 477-491.
61. Ginns, J. H. 1986. Compendium of Plant Diseases and Decay Fungi in Canada, 1960-1980. Research Branch, Agriculture Canada. Publication 1813.
62. Griffin, D. H., Quinn, K. and McMillen, B. 1986. Regulation of hyphal growth rate of *Hypoxylon mammatum* by amino acids: stimulation by proline. Exper. Mycol. 10: 307-314.
63. Griffith, G. S. and Boddy, L. 1988. Fungal communities in attached ash (*Fraxinus excelsior*) twigs. Trans. Br. Mycol. Soc. 91: 599-606.
64. Griffith, G. S. and Boddy, L. 1990. Fungal decomposition of attached angiosperm twigs. I. Decay community development in ash, beech and oak. New Phytol. 407-415.
65. Harrington, T. C. 1992. *Leptographium*. Pages 129-133 in: Methods for Research on Soilborne Phytopathogenic Fungi. L. L. Singleton, J. D. Mihail and C. M. Rush, eds. American Phytopathological Society, St. Paul, MN.
66. Hawksworth, D. L. 1991. The fungal dimension of biodiversity: magnitude, significance, and conservation. Mycol. Res. 95: 641-655.
67. Hawksworth, D. L. and Colwell, R. R. 1992. Report of a joint IUBS/IUMS/ workshop in support of the IUBS/SCOPE/UNESCO Programme on biodiversity. World J. Microbiol. Biotechnol. 8: ii-iv.
68. Holdenrieder, O. and Sieber, T. N. 1992. Fungal associations of serially washed healthy non-mycorrhizal roots of *Picea abies*. Mycol. Res. 96: 151-156.

69. Hyde, K. D. 1990. Study of the vertical zonation of intertidal fungi on *Rhizophora apiculata* at Kapong mangrove, Brunei. Aquatic Bot. 36: 255-262.
70. Johnston, A. and Booth, C. 1983. The Plant Pathologist's Pocketbook. 2nd Edition. CAB International, Slough. 439 pp.
71. Jones, R. J., Sizmur, K. J. and Wildman, H. G. 1991. A miniaturized system for storage of fungal cultures in water. The Mycologist 5: 184-186.
72. Jong, S. C. and Edwards, M. J. 1991. American Type Culture Collection Catalogue of Filamentous Fungi, 18th Edition. American Type Culture Collection, Rockville, MD.
73. Kendrick, B., Samuels, G. J., Webster, J. and Luttrell, E. S. 1979. Techniques for establishing connections between anamorph and teleomorph. Pages 635-652 in: The Whole Fungus. Vol. 2. B. Kendrick, ed. National Museums of Science, National Museums of Canada, and the Kananaskis Foundation, Ottawa .
74. Kendrick, W. B. and Burgess, A. 1962. Biological aspects of the decay of *Pinus sylvestris* leaf litter. Nova Hedwigia 4: 313-342.
75. Kohlmeyer, J. and Kohlmeyer, E. 1979. Marine Mycology, The Higher Fungi. Academic Press, New York. 690 pp.
76. Larson, T. G. and Nuss, D. L. 1993. Cyclophilin-dependent stimulation of transcription by cyclosporin A. Proc. Natl. Acad. Sci. USA 90: 148-152.
77. Leach, C. M. 1971. A practical guide to the effects of visible light and ultraviolet light on fungi. Pages 609-664 in: Methods in Microbiology. Vol. 4. C. Booth, ed. Academic Press, London
78. Leuchtmann, A., Petrini, O., Petrini, L. E. and Carroll, G. C. 1992. Isozyme polymorphism in six endophytic *Phyllosticta* species. Mycol. Res. 96: 287-294.
79. Leuchtmann, A. and Clay, K. 1988. *Atkinsonella hypoxylon* and *Balansia cyperi*, epiphytic members of the Balansiae. Mycologia 80: 192-199.
80. Millar, C. S. and Richards, G. M. 1974. A precautionary note on collection of plant specimens for mycological examination. Trans. Br. Mycol. Soc. 63: 607-612.
81. Minter, D. W. 1981. *Lophodermium* on pines. Mycol. Pap. 147: 1-54.
82. Mitchell, C. P., Millar, C. S. and Minter, D. W. 1978. Studies on decomposition of scots pine needles. Trans. Br. Mycol. Soc. 71: 343-348.
83. Monk, K. A. and Samuels, G. J. 1990. Mycophagy in grasshoppers (*Orthoptera*: Acrididae) in Indo-Malayan rain forests. Biotropica 22: 16-21.
84. Muhsin, T. M. and Booth, T. 1987. Fungi associated with halophytes of an inland salt marsh, Manitoba, Canada. Can. J. Bot 65: 1137-1151.

85. Muhsin, T. M. and Zwain, K. H. 1989. A fungal endophyte associated with desert parasitic plant. Kavaka 17: 1-5.
86. Nag Raj, T. R. 1981. Coelomycete systematics. Pages 43-84 in: Biology of Conidial Fungi. Vol. 1. G. T. Cole and B. Kendrick, eds. Academic Press, New York .
87. Newell, S. T. 1976. Mangrove fungi: The succession in the mycoflora of red mangrove (*Rhizophora mangle* L.). Pages 51-91 in: Recent Advances in Aquatic Mycology. E. B. G. Jones, ed. John Wiley, New York .
88. Noble, H. M., Langley, D., Sidebottom, P. J., Lane, S. J. and Fisher, P. J. 1991. An echinocandin from an endophytic *Cryptosporiopsis* sp. and *Pezicula* sp. in *Pinus sylvestris* and *Fagus sylvatica*. Mycol. Res. 95: 1439-1440.
89. Ondeyka, J., Hensens, O. D., Zink, D., Ball, R., Lingham, R. B., Bills, G., Dombrowski, A. and Goetz, M. 1992. L-696,474, a novel cytochalasin as an inhibitor of HIV-1 protease. II. Isolation and structure. J. Antibiot. 45: 679-685.
90. Ouellette, G. B. and Ward, E. W. B. 1970. Low-temperature requirements for ascospore germination and growth of *Hypoxylon diathrauston*. Can. J. Bot 48: 2223-2225.
91. Palm, M. E. 1991. Taxonomy and morphology of the synanamorphs *Pilidium concavum* and *Hainesia lythri* (Coelomycetes). Mycologia 83: 693-872.
92. Parkinson, D. and Kendrick, W. B. 1960. Investigations of soil microhabitats. Pages 22-28 in: The Ecology of Soil Fungi. D. Parkinson and J. S. Waid, eds. Liverpool University Press, Liverpool .
93. Petrini, L. E. and Petrini, O. 1985. Xylariaceous fungi as endophytes. Sydowia 38: 216-234.
94. Petrini, L. E., Petrini, O. and Laflamme, G. 1989. Recovery of endophytes of *Abies balsamea* from needles and galls of *Paradiplosis tumifex*. Phytoprotection 70: 97-103.
95. Petrini, O. 1982. Notes on some species of *Chloroscypha* endophytic in Cupressaceae of Europe and North America. Sydowia 35: 206-222.
96. Petrini, O. 1984. Endophytic fungi in British Ericaceae: a preliminary study. Trans. Br. Mycol. Soc. 83: 510-512.
97. Petrini, O. 1986. Taxonomy of endophytic fungi in aerial plant tissues. Pages 175-187 in: Microbiology of the Phyllosphere. N. J. Fokkema and J. van den Heuvel, eds. Cambridge University Press, Cambridge.
98. Petrini, O. 1987. Endophytic fungi of alpine Ericaceae. The endophytes of *Loiseleuria procumbens*. Pages 71-77 in: Arctic and Alpine Mycology II. G. A. Laursen, J. F. Ammirati and S. A. Redhead, eds. Plenum Press, New York .
99. Petrini, O. 1991. Fungal endophytes of tree leaves. Pages 179-197 in: Microbial Ecology of Leaves. J. H. Andrews and S. S. Hirano, eds. Springer-Verlag, New York .

100. Petrini, O. and Dreyfuss, M. 1981. Endophytische Pilze in epiphytischen Araceae, Bromiliaceae und Orchidaceae. Sydowia 34: 135-148.
101. Petrini, O. and Fisher, P. J. 1986. Fungal endophytes in *Salicornia perennis*. Trans. Br. Mycol. Soc. 87: 647-651.
102. Petrini, O. and Fisher, P. J. 1988. A comparative study of fungal endophytes in xylem and whole stem of *Pinus sylvestris* and *Fagus sylvatica*. Trans. Br. Mycol. Soc. 91: 233-238.
103. Petrini, O., Hake, U. and Dreyfuss, M. M. 1990. An analysis of fungal communities isolated from fruticose lichens. Mycologia 82: 444-451.
104. Petrini, O., Stone, J. and Carroll, F. E. 1982. Endophytic fungi in evergreen shrubs in western Oregon: a preliminary study. Can. J. Bot 60: 789-796.
105. Polishook, J. D., Dombrowski, A. W., Tsou, N. N., Salituro, G. M. and Curotto, J. E. 1993. Preussomerin D from the endophyte *Hormonema dematioides*. Mycologia 85: 62-64.
106. Rayner, A. D. M. and Boddy, L. 1988. Fungal Decomposition of Wood, Its Biology and Ecology. John Wiley & Sons, Chichester. 587 pp.
107. Rodrigues, K. F. 1992. Endophytic fungi in the tropical palm *Euterpe oleracea* Mart. Ph. D. Thesis. City University of New York, Lehman College. 258 p.
108. Rodrigues, K. F. and Samuels, G. J. 1990. Preliminary study of endophytic fungi in a tropical palm. Mycol. Res. 94: 827-830.
109. Rodrigues, K. F. and Samuels, G. J. 1992. *Idriella* species endophytic in palms. Mycotaxon 43: 271-276.
110. Roll-Hansen, F. and Roll-Hansen, H. 1979. Microflora of sound-looking wood in *Picea abies* stems. Eur. J. For. Pathol. 9: 308-316.
111. Roll-Hansen, F. and Roll-Hansen, H. 1980. Microorganisms which invade *Picea abies* in seasonal stem wounds. II. Ascomycetes, fungi imperfecti, and bacteria. General discussion, hymenomycetes included. Eur. J. For. Pathol. 10: 396-410.
112. Roll-Hansen, F. and Roll-Hansen, H. 1980. Microorganisms which invade *Picea abies* in seasonal stem wounds. II. General aspects. Hymenomycetes. Eur. J. For. Pathol. 10: 321-339.
113. Rossman, A. Y., Palm, M. E. and Spielman, L. J. 1987. A Literature Guide for the Identification of Plant Pathogenic Fungi. American Phytopathological Society, St. Paul, MN. 252 pp.
114. Ruscoe, Q. W. 1971. Mycoflora of living and dead leaves of *Nothofagus truncata*. Trans. Br. Mycol. Soc. 56: 463-474.
115. Sasaki, T., Takagi, M., Yaguchi, T., Miyadoh, S., Okada, T. and Koyama, M. 1992. A new antihelminthic cyclodepsipeptide, PF1022A. J. Antibiot. 45: 692-687.
116. Shear, C. L. and Stevens, N. E. 1913. The chestnut bark parasite (*Endothia parasitica*) from China. Science 38: 295-297.

117. Shear, C. L. and Stevens, N. E. 1916. The discovery of the chestnut blight parasite (*Endothia parasitica*) and other chestnut fungi in Japan. Science 43: 173-176.
118. Sherwood-Pike, M., Stone, J. K. and Carroll, G. C. 1986. *Rhabdocline parkeri*, a ubiquitous foliar endophyte of Douglas-fir. Can. J. Bot 64: 1849-1855.
119. Shigo, A. L. 1958. Fungi isolated from oak wilt trees and their effects on *Ceratocystis fagacearum*. Mycologia 50: 757-769.
120. Sieber, T. E., Sieber-Canavesi, F. and Dorworth, C. E. 1990. Identification of key pathogens of major coastal forest weeds. Canada/BC Forest Resource Development Agreement Report 113: 1-54.
121. Sieber, T. N., Sieber-Canavesi, F., Petrini, O., Ekramoddoullah, A. K. M. and Dorworth, C. E. 1991. Partial characterization of Canadian and European *Melanconium* (*Melanconis*) from some *Alnus*, *Fagus* and *Quercus* species by morphological and biochemical studies. Can. J. Bot 69: 2170-2176.
122. Sieber-Canavesi, F., Petrini, O. and Sieber, T. N. 1991. Endophytic *Leptostroma* species on *Picea abies*, *Abies alba*, and *Abies balsamea*: a cultural, biochemical and numerical study. Mycologia 83: 89-96.
123. Smith, D. and Onions, A. H. S. 1983. The Preservation and Maintenance of Living Fungi. Commonwealth Mycological Institute, Kew. 51 pp.
124. Sridhar, K. R. and Bärlocher, F. 1992. Aquatic hyphomycetes in spruce roots. Mycologia 84: 580-584.
125. Sridhar, K. R. and Bärlocher, F. 1992. Endophytic aquatic hyphomycetes of roots of spruce, birch and maple. Mycol. Res. 96: 305-308.
126. Stevenson, J. A. 1975. Fungi of Puerto Rico and the American Virgin Islands. Contribution of Reed Herbarium 23: 1-743.
127. Stipes, R. J. and Campana, R. J. (Eds). 1981. Compendium of Elm Diseases. American Phytopathological Society, St. Paul, MN. 96pp.
128. Stone, J. K. 1987. Initiation and development of latent infections by *Rhabdocline parkeri* in Douglas fir. Can. J. Bot 65: 2614-2621.
129. Stoyke, G. and Currah, R. S. 1991. Endophytic fungi from the mycorrhizae of alpine ericoid plants. Can. J. Bot 69: 347-352.
130. Subramanian, C. V. 1992. Tropical mycology and biotechnology. Curr. Sci. 63: 167-172.
131. Toti, L., Chapela, I. H. and Petrini, O. 1992. Morphometric evidence for host-specific strain formation in *Discula umbrinella*. Mycol. Res. 96: 420-424.
132. Tsao, P. H. 1970. Selective media for isolation of pathogenic fungi. Annu. Rev. Phytopathol. 8: 157-186.

133. Tscherter, H. and Dreyfuss, M. M. 1982. New metabolites. Processes for their production and their use. International Patent Application PCT/EP8/00121:
134. Tscherter, H., Hofmann, H., Ewald, R. and Dreyfuss, M. M. 1988. Antibiotic lactone compound. U.S. Patent 4,753,959.
135. Tuite, J. 1968. Plant Pathological Methods. Burgess Publishing Co., Minneapolis, MN. 239 pp.
136. Uecker, F. A. 1988. A world list of *Phomopsis* names with notes on nomenclature, morphology and biology. Mycol. Mem. 13: 1-231.
137. Uecker, F. A. 1989. A timed sequence of development of *Diaporthe phaseolorum* (Diaporthaceae) from *Stokesia laevis*. Mem. New York Bot. Gard. 49: 38-50.
138. Verhoeff, K. 1974. Latent infections by fungi. Annu. Rev. Phytopathol. 12: 99-110.
139. Wang, C. J. K. and Wilcox, H. E. 1985. New species of ectendomycorrhizal and pseudomycorrhizal fungi: *Phialophora finlandia*, *Chloridium paucisporum*, and *Phialocephala fortinii*. Mycologia 77: 951-958.
140. Warcup, J. H. 1991. The *Rhizoctonia* endophytes of *Rhizanthella* (Orchidaceae). Mycol. Res. 95: 656-659.
141. Webster, J. and Descals, E. 1981. Morphology, distribution, and ecology of conidial fungi in freshwater habitats. Pages 295-355 in: Biology of Conidial Fungi. Vol. 1. G. T. Cole and B. Kendrick eds. Academic Press, New York .
142. Wei, D. L., Chang, S. C., Lin, Y. W., Chuang, C. L. and Jong, S. C. 1992. Production of cellulolytic enzymes from the *Xylaria* and *Hypoxylon* species of the Xylariaceae. World J. Microbiol. Biotechnol. 8: 141-146.
143. Wildman, H. G. and Jones, R. J. 1991. Isolation of fungal endophytes from root samples of trees blown over at The Royal Botanic Gardens, Kew, during the 1987 hurricane. The Mycologist 5: 180-182.
144. Wildman, H. G. and Parkinson, D. 1979. Microfungal succession on living leaves of *Populus tremuloides*. Can. J. Bot 57: 2800-2811.
145. Windels, C. E. 1992. Fusarium. Pages 115-128 in: Methods for Research on Soilborne Phytopathogenic Fungi. L. L. Singleton, J. D. Milail and C. M. Rush, eds. American Phytopathological Press, St. Paul, MN.
146. Worrall, J. J. 1992. Media for selective isolation of hymenomycetes. Mycologia 83: 296-302.

CHAPTER 3

Fungal endophytes of living branch bases in several European tree species

Tadeusz Kowalski and Rolf D. Kehr

Department of Forest Pathology
Faculty of Forestry
31-425 Cracow, Poland

Institut für Pflanzenschutz im Forst
Biologische Bundesanstalt für Land- und Forstwirtschaft
D-3300 Braunschweig, Germany

Form and function of forest trees is greatly influenced by fungal colonization of bark and wood. It is a well known fact that fungi play an important role in natural pruning, i.e. colonization, decay and shedding of dead branches and thus help produce a clean bole. Therefore, research on such phenomena has an important practical aspect for forest utilization in addition to helping us understand the forest ecosystem.

In recent years, several European tree species have been examined in respect to the mycoflora of dead branches (4, 5, 6, 7, 22), but the question remained whether there is a connection between these fungi and those that colonize living branches endophytically. Therefore, the same 11 tree species already investigated in respect to dead branches (4, 5, 6, 7, 22) were examined for fungal endophytes by the authors.

The living branch base is composed of several morphologically different tissue types which are colonized to a different degree by fungi, making it important to differentiate several sample categories. After surface sterilization of six cm long basal branch segments according to the method by Sieber (31), the superficial, dead bark layer was separated from inner, living bark and from wood. The dead bark layer was termed "peridermal" and the green, living layer "subperidermal". Six samples per branch base were laid out on petri dishes. In all, 1095 branch bases were examined.

Frequency of fungal colonization was defined as the number of pieces of a given tissue type yielding at least one species in relation to the total number of pieces taken from this tissue type.

Colonization Frequency

Almost all basal segments of the investigated living branches were colonized by fungi, but there were large differences in the colonization rate of the various tissue types. The peridermal bark layer yielded at least one species in 98% of all branches, whereas subperidermal bark and wood were colonized in only 20% and 9% of all branches, respectively (Table 1).

Table 1. Frequency of fungal colonization of living branch bases (data from 23, used by permission).

	% of branches colonized		
Tree Species	Peridermal Bark	Subperidermal Bark	Wood
Abies alba Mill.	97.0	14.0	6.0
Acer pseudoplatanus L.	100.0	6.0	2.0
Alnus glutinosa L. Gaertn.	100.0	28.0	18.0
Betula pendula Roth	100.0	13.0	22.0
Carpinus betulus L.	100.0	6.3	7.5
Fagus sylvatica L.	88.1	10.6	7.5
Fraxinus excelsior L.	100.0	11.4	18.6
Larix decidua Mill.	100.0	18.2	5.5
Picea abies (L.) Karst.	100.0	43.2	6.5
Pinus sylvestris L.	99.3	27.6	4.1
Quercus robur L.	100.0	16.9	9.2
Total average	97.9	19.5	9.1

Subperidermal bark was colonized more than twice as frequently in coniferous than in deciduous trees. Vice versa was true of wood, where *Betula*, *Fraxinus* and *Alnus* showed the highest colonization rate, and *Acer* the least. A remarkably high isolation rate from the subperidermal bark layer was obtained for *Picea*.

Up to seven species per branch were isolated from dead bark, but usually branches contained one to four species in this sample category. *Abies* and *Fagus* most commonly yielded only one species, *Carpinus* and *Betula* two species. Peridermal bark of *Acer*, *Picea*, *Pinus* and *Quercus* showed a tendency to be inhabited by more species per branch than other trees, which could be due to periderm structure.

Role of Branch Diameter

Branch diameter influenced the number of fungi colonizing dead bark of a given branch, and this dependency was stronger in deciduous trees than in conifers. Branches thicker than 2 cm usually contained less species than thin branches, with the exception of *Fagus* and *Quercus*.

Table 2 illustrates the effect of branch diameter on the most commonly isolated fungi. Thin branches yielded more isolates of *Diaporthe carpini* (Fr.) Fuck., *Mollisia cinerea* (Batsch *ex* Merat) Karst., *Sclerophoma pithyophila* (Corda) Höhn., *Colpoma quercinum* (Pers. *ex* St. Am.) Wallr. and most *Phomopsis* species. *Sirodothis* spp., *Asterosporium asterospermum* (Pers. *ex* Gray) Hughes and *Fusicoccum macrosporum* Sacc. increased in frequency with rising branch diameter. Others, such as *Petrakia irregularis* van der Aa, *Pezicula cinnamomea* (DC.) Sacc. (on *Alnus*) and *Neohendersonia kickxii* (Westd.) Sutton & Pollak most commonly colonized branches of median diameter. Some individual fungi exhibited differences in this respect depending on the host. For instance, *Phialocephala* cf. *dimorphospora* (Kendrick) preferred thinner branches on *Alnus* and thicker ones on *Quercus*.

Species Diversity

Per tree species, 41 (*Acer pseudoplatanus*) to 67 (*Fagus sylvatica*) fungal taxa were isolated (Table 3). Species diversity was much higher in dead bark tissue than in subperidermal bark or wood. For instance, dead bark tissue of *Quercus robur* contained 60 species, but living bark tissue and wood yielded only 10 species and 4 species, respectively. The same species which colonized peridermal bark were often also present in living bark and wood, but their frequency in peridermal bark was usually higher than in both living bark and wood combined. *Aposphaeria* spp. are an exception in this respect and sometimes exhibited higher frequency in wood than in living bark, for instance in *Betula*, *Fraxinus*, and *Quercus*.

Although a high diversity of fungal taxa was found in dead bark, only a few of these per tree species were dominant (Table 2). Over 30% of branches were colonized by only one to three fungal species.

The density of colonization, expressed by the number of times a given fungus was present in different samples of the same branch, varied greatly. High density was observed for *Pezicula* spp., *Phomopsis* spp., *Cryptospora suffusa* (Fr.) Tul., *C. betulae* Tul., *Diaporthe carpini*, *Sirodothis* sp. and *Colpoma quercinum*. Fungi with large overall branch colonization frequency but with a low density within branches include *Coniothyrium fuckelii* Sacc., *Crumenulopsis pinicola* (Rebent.) Groves, *Lecythophora hoffmanii* (van Beyma) W. Gams & McGinnis, *Mollisia cinerea*, *Neohendersonia kickxii*, *Phialocephala* cf. *dimorphospora*, *Pseudovalsa longipes* (Tul.) Sacc., *Therrya* spp. and *Trimmatostroma betulinum* (Corda) Hughes.

Host Range

Two groups of endophytes can be differentiated in relation to the host spectrum. Some fungi are very specific and occur almost exclusively on one host, others are present on many hosts at varying degrees of frequency.

In many cases, fungi which exhibited a rather broad host range did not display a high frequency. For instance, *Alternaria alternata* (Fr.) Keissl. occurred on nine tree species, but usually only on a few branches, with the exception of *Fraxinus excelsior*. *F. excelsior* was also the only tree on which *Cladosporium cladosporioides* (Fresen.) de Vries, present on all tree species, was frequent. *Epicoccum nigrum* Link. was isolated from nine tree species, but a high percentage was found only on *Larix decidua*. Species with a wide host range are *Mollisia cinerea*, *Phialocephala* cf. *dimorphospora*, *Pezicula cinnamomea*, *P. livida* (Berk. & Br.) Rehm, *Geniculosporium serpens* Chesters & Greenhalgh and *Phomopsis* spp.

There are also a number of fungi which can be classified as host-specific. Each tree is colonized by a few such fungi, as can be seen in Table 4. Fungi were classified as host specific if they were found almost exclusively and rather frequently on only one tree species. Occasional isolations of these fungi were sometimes obtained from other trees, but only when these grew in the vicinity of the main host. Examples are *Amphiporthe leiphaemia* (Fr.) Butin, isolated from *Carpinus betulus* in a stand of *Quercus robur*; *Anthostomella pedemontana* Ferr. & Sacc., isolated from *Fagus sylvatica* under *Pinus sylvestris*; *Tubakia dryina* (Sacc.) Sutton, isolated from *Larix decidua* mixed with *Quercus robur*; *Prosthemium betulinum* Kunze *ex* Schlecht. and *Pseudovalsa lanciformis* (Fr.) Ces. & de Not, isolated from *Fraxinus excelsior* in a stand of *Betula pendula*. This supports the hypothesis that some endophytes are able to colonize morphologically similar hosts growing at the same site (25, 26). All the above mentioned fungi are known saprobes, and determination keys show that in their saprobic phase they show some kind of host specificity (1, 12, 16, 18, 24, 34).

Commonly Isolated Genera

Several genera and species of fungi were frequently encountered and seem to play a leading role in endophytic colonization of branches. Among the most important were *Mollisia* spp., *Pezicula* spp., *Phialocephala* spp.,

Phomopsis spp., *Verticicladium trifidum* Preuss. and some genera of the Xylariaceae. Their relation to substrate as saprobes is variable and not always in accordance with their host spectrum in the endophytic phase.

Mollisia species were so far only rarely recorded as endophytes (31). In this investigation, seven members of the genus were isolated, of which only *M. cinerea* showed a large host range. Its frequency was rather variable, from 2.5 % on *Carpinus* to 65.8 % on *Picea*. *M. cinerea* is known as one of the most common colonizers of decaying wood, especially *Quercus* and *Fagus* (3, 29), but as an endophyte it seems to prefer coniferous trees (Table 3). *M. cinerea* was previously isolated as a common endophyte only in needles and twigs of *Juniperus* in Switzerland (28), but on branch bases it emerged as one of the most common endophytes.

Pezicula and its *Cryptosporiopsis*-anamorphs are well known and common endophytes on many different plants and tissue types (25). In this investigation, the genus *Pezicula* was represented by five species and several further culture types which could not be attributed to known species. Only *P. cinnamomea* occurred on all tree species and ranged in frequency from 3.4% on *Pinus* to 46% on *Abies*. In literature, it is mentioned mainly on *Quercus* and *Castanea* (19, 34, 35), but evidently it possesses a large host range in the endophytic phase and is not limited to deciduous trees. This is especially underlined by the fact that *P. cinnamomea* was most common on a coniferous host, *Abies alba*, as an endophyte of the branch base. However, taxonomic work in progress may show that the isolates from coniferous hosts differ from those derived from deciduous hosts, in spite of the fact that there are no morphological differences in culture. *P. livida* was the second most frequent species and was isolated mainly from conifers, especially *Pinus*, but also with low frequency from several deciduous trees. *P. livida* has also been found in living leaves and branches of *Fagus sylvatica* (32). However, fruit bodies of *P. livida* are known only from conifers (12, 35). In respect to frequency of colonization, this is in accordance with the endophytic behavior demonstrated on branches. Another common species, *P. carpinea* (Pers.) Tul., occurred on six deciduous tree species.

Phialocephala cf. *dimorphospora* was the most frequent of the five isolated *Phialocephala* species. It occurred on all examined tree species with a frequency ranging from 2.1 % (*Pinus*) to 40 % (*Acer*) of branches. On *Picea abies* a different species of *Phialocephala* was more frequent (Kowalski and Kehr unpubl.). The *Ph.* cf. *dimorphospora* found in this investigation is identical to the one isolated from dead branches and termed *Phialocephala* sp. (4, 5, 6) and *Phialocephala dimorphospora* (7, 22) respectively, but it shows considerable cultural differences to an isolate of *Ph. dimorphospora* from CBS (Baarn). *Phialocephala* species were so far not mentioned as endophytes, but it is possible that they have been isolated by other authors. Most isolates sporulate only after long incubation periods at high humidity and low temperature (20), so they may sometimes be included under "sterile mycelia" in other investigations.

Phomopsis species were found on all tree species except *Betula*, but their frequency was very variable, ranging from 2.1% (*Pinus*) to 51.4% (*Fraxinus*). On *Fraxinus*, *Phomopsis* species were the most common taxa, followed by *Pezicula* species (Table 3). Host specificity of *Phomopsis* is hard to determine because of insufficient taxonomic treatment. Therefore, most keys delineate the species by substrate.

Several genera of the Xylariaceae were isolated. Most of them had to be identified by the anamorph or by specific structures arising in sterile cultures. An exception was *Anthostomella*, which produced fruit bodies. The *Geniculosporium*-anamorph of *Hypoxylon serpens* (Pers.) Fr. was the only Xylariaceae which occurred on all tree species. A broad spectrum was also established for the *Nodulisporium*-anamorph of *Hypoxylon fragiforme* (Pers.:Fr.) Kickx.

Petrini & Fisher (26) isolated *Verticicladium trifidum* to a high degree for *Pinus* and thus classified it as a host specific endophyte, but on branch bases it occurred on five deciduous and two coniferous hosts. Surprisingly, *V. trifidum* was even more common on *Fagus* (10.6% of all branches) than on *Pinus* (8.3% of all branches). Previously, it had been recorded only on dead pine needles (15, 21).

Influence of Site and Method

Geographical and local site factors apparently influence the composition and frequency of host-specific fungal species. For instance, *Colpoma quercinum* and *Amphiporthe leiphaemia*, among the most common branch endophytes of *Quercus robur* in this investigation, were not recorded in England (2, 27). *Asterosporium asterospermum*, the most common species on *Fagus sylvatica* in this study, was not recorded by Petrini & Fisher (26). On *Abies alba*, *Agyriellopsis caeruleo-atra* Höhn. was third most common in Sieber's (31) isolations but was not found at all in this investigation. On the other hand, *Mollisia cinerea* and *Phialocephala* species were absent in Switzerland (31). However, Sieber isolated *Phomopsis occulta* as the most common endophyte, which is in agreement with our results.

Differences in composition of the endophytic flora in branches of forest trees can be caused by several factors. The diversity of the plant community may greatly influence the degree of colonization by endophytes, which is illustrated by the presence of host-specific fungi on non-hosts growing in mixed stands together with the host. The amount of fungal inoculum present at a certain period of time might also influence which fungi appear dominant. Sieber (31) registered differences in this respect among three localities in Switzerland for *Picea* and *Abies*.

The exclusive examination of branch bases may also explain discrepancies between this study and those of other authors. Previous investigations were mainly concerned with branch and twig segments of different age (14, 26, 27, 31). Age, however, is only to a certain degree reflected by the diameter of the branch base, on which endophytic colonization was more or less strongly dependent in this investigation.

The isolation procedure employed could also influence results. This investigation relied on the method by Sieber (31), but differs from that described by other authors (17). Shorter superficial sterilization times may lead to higher percentage of ubiquitous fungi such as *Epicoccum nigrum*, *Cladosporium* spp. or fast growing fungi such as *Sordaria fimicola* (Rob.) Ces. & de Not. and *Verticicladium trifidum*, which can prevent detection of other fungi.

Role of Tissue State

In previous studies on branch endophytes, only bark and xylem were considered separately. In our investigation, the bark was subdivided into peridermal and subperidermal tissues, and this demonstrated that the physiological condition of the tissue is of great influence. The high overall colonization rates of living branches results mainly from the colonization of dead bark tissue. This should be noted when laying out whole cross sections of surface-sterilized twigs on media. Living bark tissue and to a greater extent xylem are colonized much more selectively by fungi, and this may be interpreted as "tissue specificity" (14, 26). Apparently, some fungi do not possess the capability to colonize living bark tissue and active xylem, but can only exist saprophytically in dead or dying outer bark cells of living branches. The physiological condition of the tissue seems to be the limiting factor here. The presence of many of these fungi as branch pruners in xylem after death of the branch support this hypothesis (4, 5, 6, 7, 22).

It seems desirable to define fungi colonizing bark and xylem of living branches and stems more accurately. The state of the tissue could be helpful in this respect. We propose the term "phellophytes" for fungi isolated only from dead bark layers, whereas the term "endophytes" should be used only for fungi also capable of colonizing living tissue. Since dead and living bark layers can be held apart visually, this approach also seems practically feasible. A differentiation of "endophytes" and "phellophytes" would be of interest in respect to the question of pathogenic behavior. Only a fungus capable of colonizing living tissue could be termed a latent pathogen (33), whereas other species might colonize only dead tissue and attack living tissue at times of reduced host vigour.

Comparison with Branch Pruning Fungi

The range of fungi found in living branch bases was compared with that from dead branches of the same eleven tree species (4, 5, 6, 7, 22). There is a great deal of correspondence both in respect to the fungal genera, the individual species and even in some cases to the isolation rate. Several species show almost exactly the same colonization frequency on living as on dead branch bases (Table 3). The "endophytic phase" seems to give these fungi an advantage in colonizing dying branches. Examples are *Amphiporthe leiphaemia*, *Aposphaeria* spp., *Colpoma quercinum*, *Cryptospora suffusa*, *C. betulae*, *Diaporthe carpini*, *Grovesiella abieticola* (Zell. & Goodd.) Morelet & Gremmen, *Pezicula* spp., *Phialocephala* spp., *Phomopsis* spp., *Sirodothis* spp., *Splanchnonema pupula* (Fr.) Kuntze, *Therrya* spp. and *Trimmatostroma betulinum*. To a certain degree this is also the case for *Mollisia cinerea* and *Lecythophora hoffmannii*. For some fungi, correspondence is even present at the branch diameter level (Table 2, comparison with 4, 5, 6, 7, 22). This is true for *Colpoma quercinum*, *Asterosporium asterospermum*, *Fusicoccum macrosporum*, *Diaporthe carpini*, and also for *Pezicula livida* and *Sirodothis* spp. on *Pinus*.

Some branch pruners, however, are not adapted to endophytic life. Fungi of this group were frequently found in wood of dead, mostly debarked branches (4, 5, 6, 7, 22) but were hardly isolated from living branches. This is exemplified by the bark colonizers *Hercospora taleola* (Fr.) E. Müller, *Aleurodiscus amorphus* and *Lachnellula calyciformis* (Batsch) Dharne and for colonizers of wood such as *Durella commutata* Fuck. and *D. atrocyanea* (Fr.) Höhn. Basidiomycetes also colonized branches only after death, with very few exceptions.

There are also examples for fungi which seem to be totally adapted to an endophytic life but are not able to colonize the branch extensively after it dies. For instance, the Xylariaceae were represented on living branches by many genera or species, even though they exhibited low frequency, with the exception of the *Geniculosporium*-anamorph of *Hypoxylon serpens* (11). On dead branches, however, representatives of this family were very rare (4, 5, 6, 7, 22). In light of the fact that members of this group are classified as xylotrophic endophytes (9, 10) and as saprobic colonizers of stumps, logs and branches (30), their absence from dead branches is surprising. A possible explanation is that these fungi require more constant moisture conditions in the form of larger branch diameters and stumps in order to become established in the succession of decay fungi.

Conclusions

Living branches of forest trees seem to be generally colonized in their basal parts by endophytic fungi. The fungi involved are almost exclusively Ascomycetes and Deuteromycetes, as in the case of other plant groups (25). In spite of the presence of many fungal species, only few per host are dominant, which corresponds with results presented by other authors (14, 26, 27, 31, 32).

A large degree of correspondence was found between these endophytes and the fungi later involved in "natural pruning" of the dead branch. With the exception of wounding, the natural death of a branch is a gradual process involving a reduction of vigour. Fungi already present as endophytes therefore are in a better position to colonize dying tissue.

There are many indications that natural pruning is a process actively enhanced by some fungi. Therefore, several of the fungi isolated from living branch bases are likely to be weak parasites. Their presence may prevent more aggressive parasites from colonizing the branch and spreading into the main stem. On *Quercus*, for example, *Pezicula cinnamomea* and *Colpoma quercinum* may contribute to death of weakened tissue, but the aggressive *Fusicoccum quercus* Oud., which causes annual canker, is hardly ever isolated as an endophyte.

Branch pruning seems to be a finely balanced ecological process involving a more or less specific endophytic flora. It provides fungi with a substrate for fructification and in turn helps the tree to shed branches while protecting the tissues of the main stem against more aggressive fungi. The relationship between endophytic fungi of the branch base and their host could therefore be interpreted as a mutualistic symbiosis (8); it may have the same importance for trunk and branch tissues that mycorrhizae have for the roots.

Literature Cited

1. Aa, van der, H.A. 1968. *Petrakia irregularis*, a new fungus species. Acta Bot. Neerl. 17:221-225.
2. Boddy, L., and Rayner, A.D.M. 1984. Fungi inhabiting oak twigs before and at fall. Trans. Br. Mycol. Soc. 82:501-505.
3. Breitenbach, J., and Kränzlin, F. 1981. Pilze der Schweiz. Vol. I, Ascomyceten. Mycologia, Luzern, 313 pp.
4. Butin, H., and Kowalski, T. 1983. Die natürliche Astreinigung und ihre biologischen Voraussetzungen I. Die Pilzflora der Buche (*Fagus sylvatica*). Eur. J. For. Path. 13:322-334.
5. Butin, H., and Kowalski, T. 1983. Die natürliche Astreinigung und ihre biologischen Voraussetzungen II. Die Pilzflora der Stieleiche (*Quercus*

robur L.). Eur. J. For. Path. 13:428-439.
6. Butin, H., and Kowalski, T. 1986. Die natürliche Astreinigung und ihre biologischen Voraussetzungen III. Die Pilzflora an Ahorn, Erle, Birke, Hainbuche und Esche. Eur. J. For. Path. 16:129-138.
7. Butin, H., and Kowalski, T. 1990. Die natürliche Astreinigung und ihre biologischen Voraussetzungen V. Die Pilzflora von Fichte, Kiefer und Lärche. Eur. J. For. Path. 20:44-54.
8. Carroll, C.C. 1988. Fungal endophytes in stems and leaves: From latent pathogen to mutualistic symbiont. Ecology 69:2-9.
9. Chapela, I.H. 1989. Fungi in healthy stems and branches of American beech and aspen; a comparative study. New Phytologist 113:65-75.
10. Chapela, I.H., and Boddy, L. 1988: Fungal colonization of attached beech branches. New Phytologist 110:47-57.
11. Chesters, C.G.C., and Greenhalgh, G.N. 1964. *Geniculosporium serpens* gen. et sp. nov., the imperfect state of *Hypoxylon serpens*. Trans. Br. Myc. Soc. 47: 393-401.
12. Dennis, R.W.G. 1978. British Ascomycetes. J. Cramer, Vaduz, 585 pp.
13. Diedicke, H. 1911. Die Gattung *Phomopsis*. Annal. Mycol. 9:8-35.
14. Fisher, P.J., and Petrini, O. 1990. A comparative study of fungal endophytes in xylem and bark of *Alnus* species in England and Switzerland. Mycol. Res. 94:313-319.
15. Gremmen, J. 1960. A contribution to the mycoflora of pine forests in The Netherlands. Nova Hedw. 1:251-288.
16. Gremmen, J., and Morelet, M. 1971. A propos de *Grovesiella abieticola* (Zell. et Goodd.) Morelet et Gremmen. Eur. J. For. Path. 1: 80-87.
17. Griffith, G.S., and Boddy, L. 1988. Fungal communities in attached Ash (*Fraxinus excelsior*) twigs. Trans. Br. Mycol. Soc. 91:599-606.
18. Grove, W.B. 1935/37. British Stem and Leaf Fungi. Vol. I and II. Cambridge Univ. Press.
19. Johansen, G. 1949. The Danish species of the Discomycete genus *Pezicula*. Dansk Bot. Ark. 13: 1-26.
20. Kendrick, W.B. 1961. The *Leptographium* complex: *Phialocephala* gen. nov. Can. J. Bot. 39:1079-1085.
21. Kowalski, T. 1988. Zur Pilzflora toter Kiefernnadeln. Z. Mykologie 54:159-173.
22. Kowalski, T., and Butin, H. 1989. The natural pruning of branches and its biological conditions IV. The fungal flora of fir (*Abies alba* Mill.). Z. Mykol. 55:189-196.
23. Kowalski, T., and Kehr, R.D. 1992: Endophytic fungal colonization of branch bases in several forest tree species. Sydowia 43:137-168.
24. Munk, H. 1957. Danish Pyrenomycetes. Munksgaard, Kopenhagen, 491 pp.

25. Petrini, O. 1986. Taxonomy of endophytic fungi of aerial plant tissues. Pages 175-187 in: Microbiology of the Phyllosphere. N. J. Fokkema, and J. van den Heuvel, eds. Cambridge University Press.
26. Petrini, O., and Fisher, P.J. 1988. A comparative study of fungal endophytes in xylem and whole stem of *Pinus sylvestris* and *Fagus sylvatica*. Trans. Br. Mycol. Soc. 91:233-238.
27. Petrini, O., and Fisher, P.J. 1990. Occurrence of fungal endophytes in twigs of *Salix fragilis* and *Quercus robur*. Mycol. Res. 94:1077-1080.
28. Petrini, O., and Müller, E. 1979. Pilzliche Endophyten, am Beispiel von *Juniperus communis* L.. Sydowia Ann. Mycol. (Ser. II) 32:224-249.
29. Rehm, H. 1986. Die Pilze Deutschlands, Õsterreichs und der Schweiz. Vol. III. Hysteriaceen und Discomyceten. In: Rabenhorst (ed.). Kryptogamen-Flora.- Eduard Kummer, Leipzig. 1272 pp.
30. Rogers, J.D., and Callan, B.E. 1986. *Xylaria polymorpha* and its allies in continental United States. Mycologia 78:391-400.
31. Sieber, T.N. 1989. Endophytic fungi in twigs of healthy and diseased Norway spruce and white fir. Mycol. Res. 92:322-326.
32. Sieber, T., and Hugentobler, C. 1987. Endophytische Pilze in Blättern und Ästen gesunder und geschädigter Buchen (*Fagus sylvatica* L.). Eur. J. For. Path. 17:411-425.
33. Sinclair, J.B. 1991. Latent infection of soybean plants and seeds by fungi. Plant Disease 75:220-224.
34. Sutton, B.C. 1980. The Coelomycetes. CAB, Surrey, Kew. 696 pp.
35. Wollenweber, H.W. 1939. Discomyzetenstudien (*Pezicula* Tul. und *Ocellaria* Tul.). Arb. Biol. Reichsanst. Land.- u. Forstw. 22:521-571.

Table 2: Most common fungal taxa in relation to branch diameter (data from 23, used by permission; for fungal authors, see Table 3).

Tree Species	Fungal Taxa	% of branches colonized diameter (cm)		
		< 1	1.1-2	>2
Abies alba	*Grovesiella abieticola*	19.6	27.3	28.6
	Pezicula spp.	65.2	69.7	33.3
	Phomopsis spp.	73.9	48.5	0
Acer pseudoplatanus	*Petrakia irregularis*	21.1	82.4	42.8
	Phomopsis spp.	78.9	29.4	7.1
Alnus glutinosa	*Cryptospora suffusa*	84.2	77.8	69.2
	Pezicula cinnamomea	15.8	50.0	15.4
	Phialocephala cf. *dimorphospora*	26.3	33.3	23.1
Betula pendula	*Cryptospora betulae*	85.3	92.0	87.8
	Pseudovalsa lanciformis	26.5	40.0	48.8
Carpinus betulus	*Diaporthe carpini*	51.5	17.2	11.1
	Pezicula spp.	60.6	69.0	83.3
Fagus sylvatica	*Asterosporium asterospermum*	26.2	33.3	48.7
	Neohendersonia kickxii	11.9	30.9	10.5
	Pezicula spp.	11.9	21.4	27.6
	Fusicoccum macrosporum	7.1	4.8	25
Fraxinus excelsior	*Phomopsis* spp.	48.0	52.4	54.2
	Pezicula cinnamomea	16.0	38.1	29.2
Larix decidua	*Sirodothis* spp.	29.2	53.3	87.5
	Phialocephala cf. *dimorphospora*	41.7	46.7	12.5
Picea abies	*Mollisia cinerea*	77.9	61.9	54.3
	Pezicula livida	47.1	46.5	58.7
	Pezicula cinnamomea	20.6	39.0	54.3
Pinus silvestris	*Pezicula livida*	65.9	75.7	71.9
	Sclerophoma pityophila	54.5	21.6	14.1
	Sirodothis spp.	13.6	18.9	28.9
Quercus robur	*Amphiporthe leiphaemia*	45.2	48.9	46.5
	Colpoma quercinum	95.2	62.2	60.5
	Pezicula cinnamomea	11.9	20.0	60.5
	Phialocephala cf. *dimorphospora*	14.3	15.6	27.9

Table 3. Fungal endophytes of the branch base (data from 23, used by permission). Explanations: a) P = peridermal bark, S = subperidermal bark, W = wood, b) WD = wood of dead branches; see 4, 5, 6, 7, 22, according to the following scale: + = on up to 5 % of dead branches, ++ = on 6-20 % of dead branches, +++ = on 21-40 % of dead branches, ++++ = on over 40 % of dead branches c) only anamorph observed d) isolated only from one or two branches; see Table 6b in 23.

Species	% of living branches colonized			% of dead branches colonized
	P[a]	S[a]	W[a]	WD[d]
ABIES ALBA				
Coniothyrium fuckelii Sacc.	3.0			
Didymosphaeria igniaria Booth	4.0			
Durandiella gallica Morelet[c]	6.0	++		
Geniculosporium serpens Chesters & Greenhalgh	12.0			
Godronia cassandrae Peck	5.0			
Grovesiella abieticola (Zell. & Goodd.) Morelet & Gremmen[c]	24.0	++		
Lecytophora hoffmannii (van Beyma) W. Gams & McGinnis	7.0	2.0	++	
Mollisia cinerea (Batsch:Fr.) Karst.	12.0	++		
Pezicula cinnamomea (DC.) Sacc.	46.0			
Pezicula livida (Berk. & Br.) Rehm	14.0	+++		
Phialocephala cf. *dimorphospora* Kendrick	7.0	++		
Phialocephala fortinii Wang & Wilcox	6.0	++++		
Phialocephala sp.	3.0	+++		
Phomopsis spp.	50.0	4.0	++	
Rosellinia sp.	3.0			
Sclerophoma pithyophila (Corda) Höhn.	3.0	+		
Torula sp.	5.0			
sterile mycelia	6.0	1.0	1.0	
others[d]	33.0	1.0	6.0	
ACER PSEUDOPLATANUS				
Aposphaeria spp.	10.0	2.0	2.0	+++
Diplodina acerina (Pass.) Sutton	22.0	2.0		
Geniculosporium serpens Chesters & Greenhalgh	8.0			
Godronia urceolus (Alb. *ex* Schw.) Karst.	6.0			
Mollisia cinerea (Batsch *ex* Merat) Karst.	18.0			
Mollisia sp.	12.0			
Myxosporium carneum Lib.	6.0			

Table 3 (cont.)

Petrakia irregularis van der Aa	48.0			
Pezicula acericola (Peck) Sacc.	6.0	++		
Pezicula carpinea (Pers.) Tul.	6.0			
Pezicula cinnamomea (DC.) Sacc.	30.0	2.0		
Phialocephala cf. *dimorphospora* Kendrick	40.0	+++		
Phomopsis pustulata Died.	42.0	++		
Splanchnonema pupula (Fr.) Kuntze[c]	36.0	+++		
Torula sp.	16.0			
sterile mycelia	8.0			
others	56.0			
ALNUS GLUTINOSA				
Aposphaeria spp.	12.0	6.0	6.0	+++
Coniochaeta velutina (Fuck.) Munk				
Cryptospora suffusa (Fr.) Tul.	78.0	6.0	2.0	+++
Melanconis thelebola (Fr.) Saccc	6.0			
Melanconium apiocarpum Link	8.0	2.0	2.0	
Mollisia cinerea (Batsch *ex* Merát) Karst.	6.0	++		
Pezicula alni Rehm[c]	6.0			
Pezicula cinnamomea (DC.) Sacc.	28.0	4.0	+	
Pezicula cf. *carpinea* (Pers.) Tul.	6.0			
Phialocephala cf. *dimorphospora* Kendrick	24.0	++++		
Phialocephala sp.	6.0			
Phialophora spp.	8.0	2.0	+	
Phoma sp.	8.0			
Phomopsis alnea Höhn.	10.0			
Prosthemium stellare Riess	6.0			
Tympanis alnea (Pers.) Fr.	8.0	4.0	2.0	+
Verticicladium trifidum Preuss	6.0			
sterile mycelia	18.0			
others	48.0	6.0	4.0	
BETULA PENDULA				
Aposphaeria spp.	14.0	4.0	10.0	++
Aureobasidium pullulans (de Bary) Arn.	3.0	2.0	4.0	
Coryneum depressum Schmidt *ex* Steudel	3.0			
Cryptospora betulae Tul.	88.0	3.0	2.0	++
Geniculosporium serpens Chesters & Greenhalgh	4.0			
Gnomonia sp.	4.0			
Melanconis stilbostoma (Fr.) Tul.	16.0	+		
Mollisia cinerea (Batsch *ex* Merat) Karst.	9.0			
Petrakia irregularis van der Aa	4.0			
Pezicula cinnamomea (DC.) Sacc.	23.0	++		
Phialocephala cf. *dimorphospora* Kendrick	14.0	++++		
Pleomassaria siparia (Berk. & Br.) Sacc.	7.0	1.0		

Table 3 (cont.)

Pseudovalsa lanciformis (Fr:) Ces. & de Not.	39.0	3.0	++	
Trimmatostroma betulae (Corda) Hughes	19.0	2.0	1.0	++++
sterile mycelia	13.0	2.0		
others[d]	22.0	7.0		
CARPINUS BETULUS				
Amphiporthe leiphaemia (Fr.) Butin	3.8			
Aposphaeria spp.	12.5	1.3	+++	
Cryptospora suffusa (Fr.) Tul.	6.3			
Daldinia sp.	5.0	2.5		
Diaporthe carpini (Fr.) Fuck.	30.0	2.5	++++	
Geniculosporium serpens Chesters & Greenhalgh	13.8			
Hypoxylon cf. *fragiforme* (Pers.:Fr.) Kickx	3.8			
Melanconiella spodiaea (Tul.) Sacc.	13.8	++		
Pezicula carpinea (Pers.) Tul.	51.3	2.5	++	
Pezicula cinnamomea (DC.) Sacc.	5.0			
Pezicula livida (Berk. & Br.) Rehm	3.8			
Pezicula sp.	8.8			
Phialocephala cf. *dimorphospora* Kendrick	10.0	+++		
Phomopsis sordida (Sacc.) Höhn.	7.5			
Pseudovalsa lanciformis (Fr.) Ces. & de Not.	5.0			
Verticicladium trifidum Preuss	5.0			
Xylaria spp.	6.3			
sterile mycelia	20.0			
others	35.0	1.3	3.8	
FAGUS SYLVATICA				
Anthostomella pedemontana Ferr. & Sacc.	1.9	1		
Apiognomonia errabunda (Rob. ex Desm.) Höhn.	3.8	1.3	1.3	
Aposphaeria spp.	6.9	0.6	0.6	++
Aspergillus sp.	3.1	0.6		
Asterosporium asterospermum (Pers. ex Gray) Hughes	38.8	1.3	++++	
Coryneum cf. *brachyurum* Link	2.5			
Cryptospora betulae Tul.	3.1			
Diaporthe eres Nitschke	2.5			
Fusicoccum galericulatum Sacc.	2.5	++		
Fusicoccum macrosporum Sacc.	15.0	0.6	++	
Geniculosporium serpens Chesters & Greenhalgh	11.9			
Hypoxylon deustum (Hoffm.:Fr.) Grev.	1.9			
Hypoxylon fragiforme (Pers.:Fr.) Kickx	2.5			

Table 3 (cont.)

Lecythophora hoffmannii (van Beyma) W. Gams & McGinnis	5.0		
Mollisia cinerea (Batsch *ex* Merat) Karst.	3.8	0.6	
Neohendersonia Kickxii (Westd.) Sutton & Pollak	16.3	+++	
Pezicula carpinea (Pers.) Tul.	6.9	+++	
Pezicula cinnamomea (DC.) Sacc.	15.0		
Pezicula sp.	3.1		
Phialocephala cf. *dimorphospora* Kendrick	5.0	+++	
Trimmatostroma betulinum (Corda) Hughes	1.9	1.3	
Verticicladium trifidum Preuss	10.6		
Xylaria spp.[c]	10.6	1.3	
sterile mycelia	7.5	1.3	
others	17.5	3.1	4.4
FRAXINUS EXCELSIOR			
Alternaria alternata (Fr.) Keissl.	22.9	+	
Cladosporium cladosporioides (Fresen.) de Vries	4.3		
Coniothyrium fraxini (Died.) Pandr. & Syd.	7.1		
Coniothyrium fuckelii Sacc.	4.3	+++	
Aposphaeria sp.	1.4	5.7	+++
Cyclothyrium juglandis (Schum. ex Rabenh.) Sutton	7.1	2.9	2.9
Fusarium spp.	5.7	+++	
Gelatinosporium cf. *betulinum* Peck	10.0		
Geniculosporium serpens Chesters & Greenhalgh	7.1		
Mollisia cinerea (Batsch *ex* Merat) Karst.	11.4	1.4	
Phialocephala cf. *dimorphospora* Kendrick	18.6	+	
Phialophora sp.	4.3		
Pezicula cf. *carpinea* (Pers.) Tul.	5.7		
Pezicula cinnamomea (DC.) Sacc.	27.1	1.4	
Phomopsis spp.	51.4	++++	
Pseudovalsa lanciformis (Fr.) Ces. & de Not[c]	8.6	1.4	
Ulocladium cf. *consortiale* (Thüm.) Simmons	5.7		
Xylohypha sp.	25.7	2.9	1.4
sterile mycelia	17.1		
others	42.9	2.9	7.1
LARIX DECIDUA			
Alternaria alternata (Fr.) Keissl.	5.5		
Coniothyrium fuckelii Sacc.	5.5	1.8	+
Epicoccum nigrum Link	5.5	+	
Gelatinosporium cf. *betulinum* Peck	14.5	1.8	
Gelatinosporium spp.	12.7	+	

Table 3 (cont.)

Geniculosporium serpens Chesters & Greenhalgh	29.1	1.8		
Hypoxylon fragiforme (Pers.:Fr.) Kickx	5.5			
Lecythophora hoffmannii (van Beyma) W. Gams & McGinnis	5.5	++		
Mollisia cinerea (Batsch *ex* Merat) Karst.	25.5			
Pezicula cinnamomea (DC.) Sacc.	12.7	1.8		
Pezicula livida (Berk. & Br.) Rehm	7.3	++++		
Phialocephala cf. *dimorphospora* Kendrick	34.5	+++		
Phomopsis occulta Trav.	12.7	+		
Sirodothis spp.	52.7	10.9	1.8	+
Torula sp.	7.3			
Trimmatostroma scutellare (Berk. & Br.) M. B. Ellis	9.1	+		
Tubakia dryina (Sacc.) Sutton	9.1			
sterile mycelia	16.4			
others	47.3	1.8	1.8	
PICEA ABIES				
Alternaria alternata (Fr.) Keissl.	1.9	+		
Aposphaeria sp.	3.2	1.3	0.6	+
Aspergillus sp.	3.2			
Aureobasidium pullulans (de Bary) Arn.	1.3	1.9		
Cystodendron sp.	2.6	1.3		
Epicoccum nigrum Link	3.9	+		
Geniculosporium serpens Chesters & Greenhalgh	23.9			
Lecythophora hoffmannii (van Beyma) W. Gams & McGinnis	15.5	1.9	0.6	+++
Mollisia cinerea (Batsch *ex* Merat) Karst.	65.8	13.5	++	
Pezicula cinnamomea (DC.) Sacc.	35.5	2.6		
Pezicula livida (Berk. & Br.) Rehm	50.3	10.3	1.9	
Phialocephala cf. *dimorphospora* Kendrick	11.0	1.9	++++	
Phialocephala sp.	25.1	3.2	++++	
Phialophora fastigiata (Lagerb. & Melin) Conant	1.9	+		
Phomopsis occulta Trav.	12.3	++		
Phomopsis sp.	6.5			
Rhizoctonia sp.[c]	2.6			
Rosellinia sp.	1.9			
Sirodothis sp.	7.1	1.9	1.3	++
Tryblidiopsis pinastri (Pers.) Karst	7.1	1.9	++	
Xylaria sp.	1.9	0.6		
sterile mycelia	11.0	1.3		
others	1.8	0.6		

Table 3 (cont.)

PINUS SYLVESTRIS				
Anthostomella formosa Kirschst.	2.1			
Cladosporium cladosporioides (Fresen.) de Vries	3.4	+		
Coniochaeta velutina (Fuck.) Munk	2.1			
Coniothyrium fuckelii Sacc.	16.6	++		
Coniothyrium pithyophilum (Höhn.) Pandr. & Syd.	2.1			
Crumenulopsis pinicola (Rebent.) Groves	12.4	1.4	++	
Epicoccum nigrum Link	2.1	+		
Lecythophora hoffmannii (van Beyma) W. Gams & McGinnis	10.3	++++		
Mollisia cinerea (Batsch *ex* Merat) Karst.	18.6	++		
Pezicula cinnamomea (DC.) Sacc.	3.4			
Pezicula livida (Berk. & Br.) Rehm	71.0	4.8	0.7	++++
Phialocephala cf. *dimorphospora* Kendrick	2.1	+		
Phomopsis occulta Trav.	2.1			
Sclerophoma pithyophila (Corda) Höhn.	28.3	3.4	++	
Sirodothis spp.	36.5	7.6	2.1	+++
Sphaeropsis sapinea (Fr.) Dyko & Sutton	2.1	+		
Therrya spp.	19.3	4.8	++	
Verticicladium trifidum Preuss	8.3			
sterile mycelia	14.5	2.8	0.7	
others	20.0	2.8	0.7	
QUERCUS ROBUR				
Alternaria alternata (Fr.) Keissl.	4.6			
Amphiporthe leiphaemia (Fr.) Butin	46.9	3.1	++	
Apiognomonia errabunda (Rob. *ex* Desm.) Höhn.	1.5			
Aposphaeria spp.	15.4	1.5	6.9	
Colpoma quercinum (Pers. *ex* St. Am.) Wallr.	71.5	3.8	++++	
Coryneum sp.	2.3			
Cystodendron sp.	3.1			
Epicoccum nigrum Link	3.1	+		
Geniculosporium serpens Chesters & Greenhalgh	13.1			
Lecythophora hoffmannii (van Beyma) W. Gams & McGinnis	4.6			
Mollisia cinerea (Batsch *ex* Merat) Karst.	16.2			
Monodictys sp.	2.3			
Nodulisporium sp.	3.8			
Pezicula carpinea (Pers.) Tul.	3.1			
Pezicula cinnamomea (DC.) Sacc.	30.8	2.3	++	
Pezicula spp.	3.8			
Phialocephala cf. *dimorphospora* Kendrick	19.2	+++		

Table 3 (cont.)

Phomopsis quercella Died.	3.8			
Pseudovalsa longipes (Tul.) Sacc.	18.5	0.8	0.8	++
Rosellinia sp.[c]	3.1	0.8		
Ulocladium chartarum (Preuss) Simmons	4.6			
Verticicladium trifidum Preuss	6.9			
Xylaria spp.[c]	5.4			
sterile mycelia	7.7	0.8		
others	20.0	1.5	1.5	

Table 4. Host specific fungal endophytes isolated from living branch bases (data from 23, used by permission).

Tree species	Fungal species
Abies alba	*Durandiella gallica, Grovesiella abieticola*
Acer pseudoplatanus	*Splanchnonema pupula, Petrakia irregularis*
Alnus glutinosa	*Cryptospora suffusa, Melanconis thelebola, Tympanis alnea*
Betula pendula	*Cryptospora betulae, Melanconis stilbostoma, Trimmatostroma betulinum*
Carpinus betulus	*Diaporthe carpini, Melanconiella spodiaea*
Fagus sylvatica	*Asterosporium asterospermum, Fusicoccum macrosporum, Neohendersonia kickxii*
Fraxinus excelsior	*Coniothyrium fraxini*
Larix decidua	*Sirodothis* sp.
Picea abies	*Tryblidiopsis pinastri*
Pinus sylvestris	*Crumenolopsis pinicola, Therrya spp.*
Quercus robur	*Amphiporthe leiphaemia, Colpoma quercinum, Pseudovalsa longipes*

CHAPTER 4

ECOLOGICAL AND PHYSIOLOGICAL ASPECTS OF HOST-SPECIFICITY IN ENDOPHYTIC FUNGI

Orlando Petrini

Microbiology Institute
Swiss Federal Institute of Technology
ETH-Zentrum
CH-8092 Zürich, Switzerland

Colonization of plant tissues by fungi is the result of a sequence of complex steps that includes recognition of the host by the symbiont, germination of the spores, penetration of the epidermis and colonization of the appropriate tissues. These processes have been studied in detail in a large number of phytopathogenic organisms (e.g., 3, 29, 38, 65) and a large body of information is available on specific host recognition by rusts and other economically important pathogenic fungi (e.g., 3, 73). Information on the physiological and genetical aspects of the mycorrhizal symbiosis has also accumulated in the last years and research on this field is rapidly progressing (26, 28). In most phytopathogenic and mycorrhizal models studied so far host–specificity underlies most colonisation processes. In the less known symptomless endophytes, host–specificity mechanisms have recently been demonstrated to initiate in many cases the establishment of endophytic symbioses (47, 67, 69, 72).

The term 'endophyte' has been re-defined to include all organisms that, at some time of their life cycle, live symptomlessly within plant tissues (47). Endophyte associations may range from intimate contact where the fungus inhabits the intercellular spaces and xylem vessels in the plant, to more or less superficial colonization of peripheral, often dying or dead tissues such as bark layers in plants with secondary growth. For endophytes colonising bark and phellem tissues the more appropriate term "phellophytes" has recently been proposed (30). Extensive reviews have dealt with the taxonomy, the biology and the evolution of endophytes (7, 8, 16, 17, 46, 47). While a large body of information is available on pathogenic and mycorrhizal endophytes, comparatively little is known on symptomless – possibly neutral or mutualistic – endophytic symbionts of aerial plant organs and on root endophytes other than mycorrhizal fungi. Basically, symptomless endophytes can be assembled in two distinct ecological groups: the mainly clavicipitaceous systemic grass endophytes, reportedly living in a mutualistic symbiosis with their hosts, and the endophytes of trees and shrubs (including the non-clavicipitaceous grass endophytes as well).

The taxonomy of clavicipitaceous endophytes was elucidated more than forty years ago (21). Successive research has mainly concentrated on the potential benefits derived by the host from the symbiosis (19). Morgan-Jones *et al.* (41) have updated the current state of taxonomic knowledge of the grass endophytes and Clay (18, 19, 20) has suggested their use in biocontrol and against plant pests. The clavicipitaceous endophytes, however, form a particular group of closely related fungi with ecological requirements and adaptations distinct from that of other endophytes. Therefore, some of the aspects discussed here and mainly derived from the analysis of non-clavicipitaceous endophytic assemblages may apply only in part to the grass endophytes, although I am convinced that the general principles of host-specificity are common to most fungi that have adapted to life in a particular habitat, regardless of their taxonomic position or of the ecological niche occupied.

Symptomless endophytes of plants other than grasses have been known for more than seventy years (2, 37, 55), yet most investigations on endophytic fungi of trees and shrubs were carried out in the last fifteen years, since the presence of symptomless ascomycetes and deuteromycetes was demonstrated in the needles of European conifers (6). In the following years, a large amount of work dealt with the compilation of detailed fungus-host lists to determine the distribution of endophytes in the plant kingdom and it is now assumed that only rarely, if ever, phanerogams are endophyte-free (46, 54). The presence of systemic endophytes in plant groups other than grasses has not yet been demonstrated but it cannot be excluded *a priori*.

Endophytic fungi of aerial plant organs provide useful models to study mutualistic and antagonistic fungus-plant symbioses. They often reside within the plant tissues for most of their life-cycle, their presence becoming manifest as host-symptoms after triggering by the appropriate ecological or physiological stimuli (31, 47). So far, however, only scant information is available on the processes that lead to a successful colonization of the host by endophytes. In natural populations of healthy plant species a series of fungi that are generally considered very specific for the corresponding host have been consistently found in the endophyte assemblages. Examples include *Hypoxylon fragiforme* (Pers.: Fr.) Kickx in *Fagus sylvatica* L. in Europe (12) and in *Fagus grandifolia* Ehrh. in North America (10), *Discula umbrinella* (Berk. *et* Br.) Morelet (teleomorph: *Apiognomonia errabunda* (Rob.) Höhnel), the causal agent of leaf anthracnose of beech and other trees, in *Fagus sylvatica* L. in Europe (59), several species of *Leptostroma* in conifer needles (5, 44, 57) and a number of known pathogens such as the beech-bark disease agent, *Nectria coccinea*, in Britain and France, *Fusarium* spp. in some agricultural crops (31, 60) and *Leptosphaeria maculans* (Desm.) Ces et de Not. [anamorph: *Phoma lingam* (Tode: Fr.) Desm.] in rapeseed (46). Careful inspection of the host-fungus lists published up to now provides useful clues for the understanding of host specificity in endophytic fungi. In general, a large number of endophytes can be isolated from a single plant species, but only few fungal taxa are dominant in each host and can be considered specific. These fungi apparently do not colonize hosts in other plant families and their occurrence is often limited to only one or few taxonomically closely related hosts (46).

Host Specificity in Clavicipitaceous Grass Endophytes

Little is known on the basic mechanisms of host specificity displayed by many grass endophytes. Only recently experimental work has been published that sheds some lights on this subject (33, 34). Most investigations on host specific establishment of grass/endophyte symbioses rely on artificial inoculation experiments carried out by introducing mycelium in wounds inflicted to the host tissue (32) or on analyses of the distribution of isozyme genotypes on different sympatric host species (33, 34). Leuchtmann and Clay (35) have detected some variation in host range among endophytic isolates of *Epichloë typhina* (Pers.) Tul. from four host grasses and have demonstrated experimentally that infection compatibility relationships are not always reciprocal. However, surprisingly little direct evidence, if any, exists that supports data from field observations and population analyses. Additional experimental work is needed to understand host range, host specificity and recognition mechanisms in grass endophytes, as knowledge of such parameters is crucial for potential applications of grass endophytes in biological control.

Host Specific Assemblages of Non-Clavicipitaceous Endophytes

From early floristic, descriptive studies, it was evident that a certain degree of specificity was involved in the establishment of endophytes in plant populations. Thus, species lists obtained for a given host were found to be more or less characteristic, with usually few strictly host specific taxa. A considerable overlap, however, could be observed between endophyte lists from different hosts, particularly from taxonomically related ones (46). In the last few years, community ordination analyses have shown that endophyte communities are usually specific at the host species level. Ericaceous species growing at the same site are generally colonized by distinct endophyte communities (45). In a study on endophytic fungi of twigs of *Pinus sylvestris* L. and *Fagus sylvatica* growing at the same site, distinct fungal communities colonized the two hosts (48). These results were later confirmed on a different model (49). Additional work on other plant species (1, 5, 59, 61) provided additional evidence for the hypothesis that each plant species develops a highly specific endophytic community. Chapela (pers. comm.) was able to prove this by simultaneously sampling endophytes from fir trees (*Abies alba* Miller) and mistletoes (*Viscum album* L.) growing on them. Even though the two plant species were not physically separated by more than one centimeter throughout their life-spans and were therefore exposed to virtually the same inoculum, the two endophytic communities turned out to be entirely different, with an overlap of less than ten fungal species out of some 70 isolated.

Host Species Specificity by Endophytic Fungi

In understanding specificity phenomena, two broad aspects must be considered, namely establishment specificity, where an endophyte specifically colonizes only selected plant species, and expression specificity, with colonization of several hosts by a given fungus that forms specific structures, usually fruiting bodies, only on a restricted number of plant taxa.

Establishment Specificity

Methods used to study establishment specificity of endophytes include ecological investigations of endophyte communities by isolation and census work, direct spore measurements and other morphological studies, water soluble protein and isozyme electrophoresis, and, more recently, DNA analysis, as well as infection and germination experiments.

Often a large number of fungal taxa can be recovered from the tissues of a single host species (1, 4, 46, 60, 63). In all plant species investigated, however, a distinct pattern of dominance by only one or few fungal taxa can be detected. For instance, *Lophodermium piceae* (Fuckel) Höhn. is the most frequent endophyte of *Picea abies* Karst. (4, 57). *Phyllosticta multicorniculata* Bissett *et* Palm is the most important colonizer of balsam fir needles [*Abies balsamea* (L.) Mill.] (44). Leaves of beech in Europe are mainly colonized by *Discula umbrinella* (59). *Hypoxylon fragiforme* was consistently isolated from healthy beech trees in Europe (10, 12) and *H. mammatum* (Wahl.) Miller was very frequent, although not abundant, in aspen in the northeastern U.S. (10). The presence of host-specific endophytes in red alder (*Alnus rubra* Bong.) has been demonstrated in a study designed to isolate potentially pathogenic endophytes to be used in biocontrol programs (61).

Cultural characters and substrate utilization tests are often useful to characterize endophytes. The growth rates on different media of morphologically indistinguishable *Leptostroma* isolates correlate well with the hosts from which the strains have been derived (64). In the same study, substrate utilization tests have also allowed a discrimination of isolates in accordance with their origin.

Host specific colonization can sometimes be evidenced by simple morphometric analyses. Host related variation of ascospore size in *Hypoxylon fuscum* (Pers.: Fr.) Fr. has been demonstrated by means of classical microscopical measurements (43). Chapela (11), using a particle sizing device, has shown a similar pattern for *Hormonema* spp. isolated from mistletoe (*Viscum album*) and European white fir (*Abies alba*). Culturally indistinguishable isolates of *Discula umbrinella* can be grouped in distinct host-specific morphotypes by microscopic measurements and electronic spore sizing (69).

The usefulness of electrophoretic methods in fungal taxonomy has been demonstrated several times. Using water soluble protein electrophoresis several races can be distinguished within the conifer pathogen *Gremmeniella abietina* (Lagerberg) Morelet (52, 53). Host-related differences have been detected in the electrophoretic profiles of water soluble proteins of *Leptostroma* isolates (64). Similar results have been reported for *Melanconium* spp. from alder in Europe and Canada (62).

Almost simultaneously the advantages of ease and speed provided by multilocus enzyme electrophoresis have been utilized to detect host-related differences in endophytic isolates. Considerable variation is present within isolates of *Atkinsonella hypoxylon* (Peck) Diehl from different hosts (33). Similar results have been obtained with conifer-inhabiting *Phyllosticta* spp. (36) and with isolates of *Discula umbrinella* from beech, oak (*Quercus* spp.) and chestnut (*Castanea sativa* L.) (67). Pectic and amylolytic enzymes electrophoresis has been successfully used with *Leptostroma* isolates (64). Pectic enzymes production of *D. umbrinella* isolates from different hosts has been studied by polyacrylamide gel electrophoresis (67, 68). All isolates investigated are able to produce pectin pectyl hydrolase and polygalacturonases and the banding patterns are consistent with the host origin of the isolates.

DNA markers have now been successfully used to identify fungal species or intraspecific strain and race formation (39). DNA fingerprinting may be useful to differentiate genera of fungi (40). Recombinant DNA techniques are now widely used in fungal ecology and help substantially in the study of host-specificity. For instance, Haemmerli *et al.* (27) have studied genetic variation in 30 isolates of *D. umbrinella* derived from beech, chestnut and oak using randomly amplified polymorphic DNA (RAPD, 71) and restriction fragment length polymorphism (RFLP) markers. Cluster analysis using the data of all arbitrary primed amplified DNA-fragments have demonstrated that the isolates could be placed in groups corresponding to their host origin. RAPD marker patterns and the corresponding RFLPs are in good agreement with other physiological traits such as isozyme analysis (67) and pectinase production (67, 68).

All methods described so far provide only indirect indication for host specificity by individual endophytic strains, being based on diagnostic methods rather than on the direct observation of the host-fungus interactions.

More direct evidence can be obtained from germination and infection experiments. Establishment specificity is probably mediated by finely-tuned recognition and signalling between host and fungus. In the last years we have been addressing this question in two selected endophyte/host pairs, viz. in the *Hypoxylon fragiforme/Fagus sylvatica* and *Discula umbrinella/Fagus sylvatica* models.

Early recognition of a potential host by *Hypoxylon* species is mediated by a newly discovered spore activation mechanism (13). Ascospores of *Hypoxylon fragiforme*, believed to be the infection units, can remain viable but unresponsive to all common germination inducers for more than a year until a molecular message, derived from the host cell wall, is sensed by specific receptor(s) on the spore. Within minutes of activation by the appropriate signal, ascospores undergo dramatic changes, collectively called eclosion (13). Chapela *et al.* (14) have extracted a factor from living, healthy beech bark, which induces eclosion and germ tube formation in *H. fragiforme*. The specificity of the eclosion factor is underlined by experiments with heterologous (unusual) host plants. Ascospores from *H. fragiforme* were confronted with leaf squares from 16 vascular plants. The response of ascospores to the presence of leaf squares in the incubation medium was either a very high eclosion and germination rate or hardly any germination at all. Only a few of the tested plant species, including beech, the usual host of the fungus, elicited high eclosion rates. This suggests that several plants produce a characteristic signature that can be differentially interpreted by different fungi as a cue to germination. This signature is a mixture of two cis-monolignols, faguside and syringoside, that are present on the host surface and are apparently responsible for the induction of eclosion and germ tube formation in *H. fragiforme* and other members of the Xylariaceae (14). Eclosion is a two–step process (15). In the first step the activated spore body is released from the exosporium. This phase is probably mediated by a Na^+–sensitive receptor to host-derived monolignol glucosides, although potential mechanical energy in the inner wall layers may also be released upon activation of the putative receptor (15). In the second step the released spore body swells and exposes the inner cell wall layer. In contrast to the dormant ascospore, the spore surface exposed upon eclosion binds readily both chitin and glucose/mannose-specific lectins.

Specificity phenomena have also been studied in beech isolates of *Discula umbrinella*. In this model, the ability of conidia to attach to different host and non-host surfaces was first tested by applying fresh, autoclaved, and cycloheximide-treated conidia to the leaf surface of beech and of the two taxonomically closely related chestnut and oak, as well as to non-host plant and artificial surfaces, and by subsequently attempting to remove the conidia by washing the material under running water (70). On beech leaves, the attachment gradually increased over a twenty-four hour period, with a maximum after 16 to 24h. The adhesion of conidia to the host surface was clearly pH-dependent and reached an optimum at pH 5, to decline at higher pH values. Heat and chemical treatment significantly reduced attachment. A high proportion of fresh conidia on beech leaves was resistant to removal, whereas only a small amount remained attached to the other hosts. Almost no binding was detected on non-host and artificial surfaces. Autoclaved and cycloheximide-treated conidia adhered to beech leaves, even if in lower percentages than the untreated ones, but were easily washed off from the other surfaces.

In a second set of experiments, the involvement of surface sugars in the attachment of *D. umbrinella* conidia to the host surface and the role of the conidial sheath in the recognition and adhesion process was investigated (67, 72). Treatment of the conidial surface with different lectins effectively inhibited the adhesion of conidia to the host. Enzymatic ablation or chemical treatment of the spores did not alter the binding patterns of the glucose/mannose specific Concanavalin A, but resulted in a strong reduction of the attachment and changes in the fluorescence patterns after treatment with the N-acetyl-glucosamine specific wheat germ agglutinin. Transmission electron microscopy revealed that digestion with glucanases and other polysaccharide hydrolases modified only slightly the structure of the conidial sheath, while proteases completely dissolved it, leaving a comparatively smooth cell wall. After protease treatment, the adhesion was completely inhibited, suggesting that proteins, possibly enzymes, are also involved in the attachment process.

Petrini *et al.* (51) postulated that host-specific strains may evolve within endophytic fungal species. The morphometric and physiological differences among conidia of *D. umbrinella* isolates derived from different hosts may be interpreted as an indication of a recent divergence of host–specific biotypes from a common ancestor (69). The formation of host–specific strains within endophytes has now been evidenced in a number of different models (34, 62, 64). In *H. fragiforme* and *D. umbrinella* host–specific strain formation appears to be directly related to the recognition of the host by the fungus. In both models minimal quantitative differences in the cell wall composition of one of the two symbionts are responsible for the successful establishment of the symbiosis. While the host cell wall is apparently triggering the recognition processes in the *H. fragiforme*/beech symbiosis, the differential attachment in *D. umbrinella* to various hosts may be, at least partly, related to the different composition, distribution or arrangement of surface sugars on the conidial wall. Similar mechanisms may basically be responsible for the establishment of a large number of other neutral, antagonistic or mutualistic symbioses among fungal symbionts, such as *Fusarium* spp. or *Colletotrichum* spp. known to display some kind of host-specificity.

Expression Specificity

This important component of specificity, which is mostly overlooked in traditional pathogen-plant associations, is immediately evident in endophytic systems. Probably the first direct evidence of this specificity component is the demonstration that *Hypoxylon fragiforme*, until recently only known from its saprotrophic phase, was established in an inconspicuous form within the wood of practically every single beech tree so far sampled (12). A poorly known phase in the life-history of this fungus seems to involve an intracellular form that might only be established in a limited number of potential hosts (Chapela, pers. comm.). *H. fragiforme* can be isolated from several other plants, but apparently will sporulate only on a restricted number of angiospermous trees (42, 46). In other words, while *H. fragiforme* can survive in a relatively loose kind of association with a range of plant species, its full expression to complete its life-cycle takes place in a much more restricted number of hosts. The mechanisms behind this expression specificity are still very obscure, but cell-cell recognition events and regulation of phase-shifts could be responsible. Other endophytes may have life histories analogous to that evidenced for *H. fragiforme*, with a 'latent' phase followed by a more evident expression. The establishment of the inconspicuous phase (the 'endophytic phase' proper for some authors) must involve a high degree of coordination between host and fungal cells (specificity determinants), while the switch from one phase to the other could be triggered by factors so far only obscurely defined.

Organ and Tissue Specificity

Organ specificity by endophytic fungi was first demonstrated for wheat endophytes (56) and for endophytes of Norway spruce (57, 58). Aquatic and soil root samples of *Alnus glutinosa* (L.) Gaertn. are colonized by two different endophyte populations (25). This suggests that organ specificity may be the result of the adaptation by some endophytes to the particular microecological and physiological conditions present in a given organ, as functionally both aquatic and terrestrial roots are very similar, differing only in the habitat in which they are found.

Carroll *et al.* (6) first postulated some tissue specificity by endophytes, because many of the fungi isolated from the petiole of European conifers were restricted to that part and were rarely detected in more distal portions of the needle. Similar conclusions were later drawn also from studies on other model systems (9, 50). Stone (66) showed that *Rhabdocline parkeri* and *Phyllosticta* sp. co-exist in needles of *Pseudotsuga menziesii*, with *R. parkeri* confined to epidermal and hypodermal cells and *Phyllosticta* sp. occurring intercellularly in the mesophyll. Subsequent investigations (1, 22, 23, 24, 48, 49) confirmed that many endophytic fungi show a certain degree of tissue specificity.

The Significance of Host Specificity in Endophytes

Specificity and expression determinants could theoretically help to tailor more or less committed endophytic systems that could fulfill a determined economic role. An example is the proposal that endophyte-infected grasses could be more or less committed to reclamation and turf-grass uses (19). The often very high host-specificity of some endophytes makes these organisms ideally suited for biological weed control (Dorworth and Callan, this volume).

Endophyte research is strongly reinforcing the idea that not all fungi need to be deleterious or beneficial (whatever these anthropocentric concepts might mean) for their host plant. Working with natural populations of healthy, long-lived trees, we have consistently found in their endophytic communities a series of fungi that are generally considered pathogenic for the corresponding host species. Thus, the delimitation between pathogen and innocuous endophyte becomes completely blurred, at least for endophytes of aerial plant tissues and much more so for plant populations under low unidirectional selection pressures. It is clear that among fungal endophytes there is a large number of fungi that might behave under some circumstances as pathogens, while under other conditions will remain innocuous, and probably beneficial symbionts of the host plant (47). Host-specificity by endophytes and the double-sided nature of the endophytic association must be understood before practical applications for this symbiosis can be envisaged.

Acknowledgments

I should like to thank I.H. Chapela (Basle, Switzerland), C.E. Dorworth (Victoria, B.C., Canada), and T.N. Sieber (Zürich, Switzerland) for helpful discussions and for critical reading of earlier versions of the manuscript. I.H. Chapela has kindly disclosed some of his unpublished results and has helped to perform some experiments described in this paper.

Literature Cited

1. Bertoni, M.D. and Cabral, D. 1988. Phyllosphere of *Eucalyptus viminalis*. II: distribution of endophytes. Nova Hedwigia 46: 491–502.
2. Bose, S.R. 1947. Hereditary (seed-borne) symbiosis in *Casuarina equisetifolia*. Nature (London) 159: 512–514.
3. Bourett, T.M. and Howard, R.J. 1990. In vitro development of penetration structures of the rice blast fungus *Magnaporthe grisea*. Can. J. Bot. 68: 329–342.
4. Butin, H. 1986. Endophytische Pilze in grünen Nadeln der Fichte (*Picea abies* Karst.). Z. Mykol. 52: 335–346.
5. Canavesi, F. 1987. Beziehungen zwischen Endophytischen Pilzen von *Abies alba* Mill. und den Pilzen der Nadelstreue. Dissertation ETH Nr. 8325, Swiss Federal Institute of Technology, Zürich.

6. Carroll, F.E., Müller, E., and Sutton, B.C. 1977. Preliminary studies on the incidence of needle endophytes in some European conifers. Sydowia 29: 87–103.
7. Carroll, G.C. 1986. The biology of endophytism in plants with particular reference to woody perennials. Pages 205-222 in: Microbiology of the Phyllosphere. N.J. Fokkema,. and J. van den Heuvel, eds. Cambridge University Press, Cambridge.
8. Carroll, G.C. 1988. Fungal endophytes in stems and leaves: from latent pathogen to mutualistic symbiont. Ecology 69: 2 - 9.
9. Carroll, G.C. and Carroll, F.E. 1978. Studies on the incidence of coniferous needle endophytes in the Pacific Northwest. Can. J. Bot. 56: 3034–3043.
10. Chapela, I.H. 1989. Fungi in healthy stems and branches of American beech and aspen: a comparative study. New Phytol. 113: 65–75.
11. Chapela, I.H. 1991. Spore size revisited: analysis of spore populations using an automated particle sizer. Sydowia 43: 1–14.
12. Chapela, I.H. and Boddy, L. 1988. Fungal colonization of attached beech branches I. Early stages of development of fungal communities. New Phytol. 110: 39–45.
13. Chapela, I.H., Petrini, O., and Petrini, L.E. 1990. Unusual ascospore germination in *Hypoxylon fragiforme*: first steps in the establishment of an endophytic symbiosis. Can. J. Bot. 68: 2571–2575.
14. Chapela, I.H., Petrini, O. and Hagmann, L. 1991. Monolignol glucosides as specific recognition messengers in fungus/plant symbioses. Physiol. Molec. Plant Pathol. 39: 289–298.
15. Chapela, I.H., Petrini, O. and Bielser, G. 1992. The physiology of ascospore eclosion in *Hypoxylon fragiforme*: mechanisms in the early recognition and establishment of an endophytic symbiosis. Mycol. Res. 97: 157–162.
16. Clay, K. 1986. Grass endophytes. Pages 188-204 in:. Microbiology of the Phyllosphere. Fokkema, N.J. and van den Heuvel, J., eds. Cambridge University Press, Cambridge.
17. Clay, K. 1988a. Clavicipitaceous fungal endophytes of grasses: coevolution and the change from parasitism to mutualism. Pages 79-105 in: Coevolution of Fungi with Plants and Animals. K. A. Pirozynski and D. L. Hawksworth, eds. Academic Press, London.
18. Clay, K. 1988b. Fungal endophytes of grasses: a defensive mutualism between plants and fungi. Ecology 69: 10-16.
19. Clay, K. 1989. Clavicipitaceous endophytes of grasses: their potential as biocontrol agents. Mycol. Res. 92: 1–12.
20. Clay, K. 1990. Fungal endophytes of grasses. Ann. Rev. Ecol. Syst. 21: 275–297.
21. Diehl, W.W. 1950. *Balansia* and the Balansiae in America. Agriculture Monograph 4. United States Department of Agriculture, Washington, DC.

22. Fisher, P.J. and Petrini, O. 1987. Location of fungal endophytes in tissues of *Suaeda fruticosa*: a preliminary study. Trans. Br. Mycol. Soc. 89: 246–249.
23. Fisher, P.J. and Petrini, O. 1988. Tissue specificity by fungi endophytic in *Ulex europaeus*. Sydowia 40: 46–50.
24. Fisher, P.J. and Petrini, O. 1990. A comparative study of fungal endophytes in xylem and bark of *Alnus* species in England and Switzerland. Mycol. Res. 94: 313–319.
25. Fisher, P.J., Petrini, O. and Webster, J. 1991. Aquatic hyphomycetes and other fungi in living aquatic and terrestrial roots of *Alnus glutinosa*. Mycol. Res. 95: 543–547.
26. Gianinazzi-Pearson, V. and Gianinazzi, S (eds). 1985. Physiological and Genetical Aspects of Mycorrhizae. Institut National de la Recherche Agronomique, Paris. 832 pp.
27. Hämmerli, U.A., Brändle, U.E., Petrini, O. and McDermott, J.M. 1992. Differentiation of isolates of *Discula umbrinella* (teleomorph: *Apiognomonia errabunda*) from beech, chestnut and oak using RAPD markers. Mol. Plant–Micr. Interact. 5: 479–483.
28. Harley, J.L. 1989. The significance of mycorrhiza. Mycol. Res. 92: 129 –139.
29. Jones, P. and Ayres, P.G. 1974. Rhynchosporium leaf blotch of barley studied during the subcuticular phase by electron microscopy. Physiol. Plant Pathol. 4: 229–233.
30. Kowalski, T. and Kehr, R.D. 1992. Endophytic fungal colonization of branch bases in several forest tree species. Sydowia 44: 137–168.
31. Leslie, J.F., Pearson, C.A.S., Nelson, P.E. and Tousson, T.A. 1990. *Fusarium* spp. from corn, sorghum, and soybean fields in the central and eastern United States. Phytopathology 80: 343–350.
32. Leuchtmann, A. 1992. Systematics, distribution and host specificity of grass endophytes. Nat. Tox. 1: 150–162.
33. Leuchtmann, A. and Clay, K. 1989. Isozyme variation in the fungus *Atkinsoniella hypoxylon* within and among populations of its host grasses. Can. J. Bot. 67: 2600–2607.
34. Leuchtmann, A. and Clay, K. 1990. Isozyme variation in the *Acremonium/Epichloë* fungal endophyte complex. Phytopathology 80: 1133–1139.
35. Leuchtmann, A. and Clay, K. 1993. Nonreciprocal compatibility between *Epichloë typhina* and four host grasses. Mycologia 85: 157–163.
36. Leuchtmann, A., Petrini, O., Petrini, L.E. and Carroll, G.C. 1992. Isozyme polymorphism in six endophytic *Phyllosticta* species. Mycol. Res. 96: 287–294.
37. Lewis, F.J. 1924. An endotrophic fungus in the Coniferae. Nature (London) 114: 860.

38. Li, A. and Heath, M.C. 1990. Effect of intercellular washing fluids on the interactions between bean plants and fungi nonpathogenic on beans. Can. J. Bot. 68: 934–939.
39. Metzenberg, R.L. 1991. Benefactor's lecture: the impact of molecular biology on mycology. Mycol. Res. 95: 9–13.
40. Meyer, W., Koch, A., Niemann, C., Beyermann, B., Epplen, J.T. and Börner, T. 1991. Differentiation of species and strains among filamentous fungi by DNA fingerprinting. Curr. Genet. 19: 239–242.
41. Morgan-Jones, G., Phelps, R.A. and White, J.F. 1992. Systematic and biological studies in the Balansieae and related anamorphs. I. Prologue. Mycotaxon 43: 401–415.
42. Petrini, L. and Petrini, O. 1985. Xylariaceous fungi as endophytes. Sydowia 38: 216–234.
43. Petrini, L.E., Petrini, O. and Sieber, T.N. 1987. Host specificity of *Hypoxylon fuscum*: a statistical approach to the problem. Sydowia 40: 227–234.
44. Petrini, L. E., Petrini, O., and Laflamme, G. 1989. Recovery of endophytes of *Abies balsamea* from needles and galls of *Paradiplosis tumifex*. Phytoprotection 70: 97–103.
45. Petrini, O. 1985. Wirtsspezifität endophytischer Pilze bei einheimischen Ericaceae. Bot. Helv. 95: 213–238.
46. Petrini, O. 1986. Taxonomy of endophytic fungi of aerial plant tissues. Pages 175–187 in: Microbiology of the Phyllosphere. Fokkema, N.J. and van den Heuvel, J., eds. Cambridge University Press, Cambridge.
47. Petrini, O. 1991. Fungal endophytes of tree leaves. Pages 179–197 in: Microbial Ecology of the Leaves. Andrews, J.H. and Hirano S.S., eds. Springer Verlag, New York.
48. Petrini, O. and Fisher, P.J. 1988. A comparative study of fungal endophytes in xylem and whole stems of *Pinus sylvestris* and *Fagus sylvatica*. Trans. Br. Mycol. Soc. 91: 233–238.
49. Petrini, O. and Fisher, P.J. 1990. Occurrence of fungal endophytes in twigs of *Salix fragilis* and *Quercus robur*. Mycol. Res. 94: 1077–1080.
50. Petrini, O. and Müller, E. 1979. Pilzliche Endophyten am Beispiel von *Juniperus communis* L. Sydowia 32: 224–251.
51. Petrini, O., Stone, J. and Carroll, F.E. 1982. Endophytic fungi in evergreen shrubs in Western Oregon: a preliminary study. Can. J. Bot. 60: 789–796.
52. Petrini, O., Petrini, L.E., Laflamme, G. and Ouellette, G.B. 1989. Taxonomic position of *Gremmeniella abietina* and related species: a reappraisal. Can. J. Bot. 67: 2805–2814.
53. Petrini, O., Toti, L., Petrini, L.E. and Heiniger, U. 1990. *Gremmeniella abietina* and *G. laricina* in Europe: characterisation and identification of isolates and laboratory strains by soluble protein electrophoresis. Can. J. Bot. 68: 2629–2635.

54. Petrini, O., Sieber, T.N., Toti, L. and Viret, O. 1992. Ecology, metabolite production, and substrate utilization in endophytic fungi. Nat. Tox. 1: 185–196.
55. Schüepp, H. 1961. Untersuchungen über *Guignardia citricarpa* Kiely, den Erreger der Schwarzfleckenkrankheit auf *Citrus*. Phytopath. Z. 40: 258–271.
56. Sieber, T.N. 1985. Endophytische Pilze von Winterweizen (*Triticum aestivum* L.). Dissertation ETH Nr. 7725, Swiss Federal Institute of Technology, Zürich, Switzerland.
57. Sieber, T.N. 1988. Endophytische Pilze in Nadeln von gesunden und geschädigten Fichten [*Picea abies* (L.) Karsten]. Eur. J. For. Pathol. 18: 321–342.
58. Sieber, T.N. 1989. Endophytic fungi in twigs of healthy and diseased Norway spruce and white fir. Mycol. Res. 92: 322–326.
59. Sieber, T.N. and Hugentobler, C. 1987. Endophytische Pilze in Blättern und Ästen gesunder und geschädigter Buchen (*Fagus sylvatica* L.). Eur. J. For. Pathol. 17: 411–425.
60. Sieber, T., Riesen, T.K., Müller, E., and Fried, P.M. 1988. Endophytic fungi in four winter wheat cultivars (*Triticum aestivum* L.) differing in resistance against *Stagonospora nodorum* (Berk.) Cast. and Germ. = *Septoria nodorum* (Berk.) Berk. J. Phytopathol. 122: 289–306.
61. Sieber, T.N., Sieber-Canavesi, F., and Dorworth, C.E. 1991a. Endophytic fungi of red alder (*Alnus rubra*) leaves and twigs in British Columbia. Can. J. Bot. 69: 407–411.
62. Sieber, T.N., Sieber-Canavesi, F., Petrini, O., Ekramoddoullah, A.K.M., and Dorworth, C.E. 1991b. Characterization of Canadian and European *Melanconium* from some *Alnus* species by morphological, cultural, and biochemical studies. Can. J. Bot. 69: 2170–2176.
63. Sieber-Canavesi, F. and Sieber, T.N. 1988. Endophytische Pilze in Tanne (*Abies alba* Mill.) – Vergleich zweier Standorte im Schweizer Mittelland (Naturwald - Aufforstung). Sydowia 40: 250–273.
64. Sieber-Canavesi, F., Petrini, O., and Sieber, T.N. 1991. Endophytic *Leptostroma* species on *Picea abies*, *Abies alba*, and *Abies balsamea*: a cultural, biochemical, and numerical study. Mycologia 83: 89–96.
65. Smereka, K.J., Machardy, W.E., and Kausch, A.P. 1987. Cellular differentiation in *Venturia inaequalis* ascospores during germination and penetration of apple leaves. Can. J. Bot. 65: 2549–2561.
66. Stone, J.K. 1986. Foliar Endophytes of *Pseudotsuga menziesii* (Mirb.) Franco. Cytology and Physiology of the Host-Endophyte Relationship. Dissertation, University of Oregon, Eugene, Oregon..
67. Toti, L. 1993. The symbiosis *Discula umbrinella* / *Fagus sylvatica*: biochemical, ecological and morphological studies of the host–endophyte relationship. Dissertation ETH No. 10097, Swiss Federal Institute of Technology, Zurich, Switzerland.

68. Toti, L., Chassin du Guerny, A., Viret, O. and Petrini, O. 1991. Host-related pectic enzyme patterns in *Discula umbrinella.* Phytopathology 81: 1248 (abstract).
69. Toti. L., Chapela, I.H. and Petrini, O. 1992a. Morphometric evidence for host-specific strain formation in *Discula umbrinella* (teleomorph: *Apiognomonia errabunda*). Mycol. Res. 96: 420–424.
70. Toti, L., Viret, O., Chapela, I.H. and Petrini, O. 1992b. Differential attachment by the conidia of the endophyte *Discula umbrinella* (Berk. and Br.) Morelet to host and non-host surfaces. New Phytol. 121: 469–475.
71. Williams, J.G.K., Kubelik, A.R., Livak, K.J., Rafalski, J.A. and Tingey, S.V. 1990. DNA polymorphisms amplified by arbitrary primers are useful as genetic markers. Nucl. Acid Res. 18: 6531–6535.
72. Viret, O. 1993. Infection of beech leaves by the endophyte *Discula umbrinella* (Berk. & Br.) Morelet [Teleomorph: *Apiognomonia errabunda* (Rob.) Höhnel]: an ultrastructural study. Dissertation ETH No. 10073, Swiss Federal Institute of Technology, Zurich, Switzerland.
73. Wood, R.K.S. and Graniti, A. (eds.) 1976. Specificity in Plant Diseases. Plenum Press, New York. 354 pp.

CHAPTER 5

COASTAL REDWOOD LEAF ENDOPHYTES: THEIR OCCURRENCE, INTERACTIONS AND RESPONSE TO HOST VOLATILE TERPENOIDS

Francisco J. Espinosa-García*, Jeanette L. Rollinger and Jean H. Langenheim

Department of Biology
University of California
Santa Cruz, CA 95064

*Current address: Centro de Ecología
Universidad Nacional Autónoma de México.
Ap. Postal 70-275. 04510, D.F. México

Fungi causing symptomless infections in aerial plant tissues are ubiquitous among plant taxa (6, 41), and they have even been suggested to be "as common as mycorrhizae" (7). These fungi have an endophytic (i.e. within the plant) stage of variable length in which they are not harmful to the host. Moreover, either many of them never become pathogenic during their lifetime or the symptomless stage is continuous until host tissue death or damage occurs (1, 6). The term "endophyte" is used here in reference to these fungi. Under these criteria, fungi with diverse ecological roles can occur endophytically as dormant saprobes, latent pathogens, or mutualists that antagonize plant enemies or stimulate host growth and its competitive ability (2, 3, 6, 9, 54). Because of the potential ecological importance of endophytes, we explored endophyte-plant relationships using coastal redwood [*Sequoia sempervirens* (D.Don *ex* Lamb.) Endl.] and its associated leaf endophytes as a study system. We have focused on the diversity and occurrence of coastal redwood endophytes, the interaction among redwood endophytes, and the relationship between host volatile terpenoids (bioactive secondary chemicals) and endophytes in redwood leaves.

Coastal redwood is a dominant tree of forests occurring in a narrow coastal band (5-35 miles wide) from central California to southern Oregon, U.S.A. (McBride and Jacobs, unpublished data). Redwood is one of the longest lived trees as well as the tallest tree in the world. It propagates sexually with seeds and vegetatively with basal sprouts (suckers), which are produced profusely when the tree is cut, the base is burned, otherwise injured or fully exposed to sunlight. Coastal redwood is unusual in that commercially important pests (herbivore or pathogen) are generally absent or entirely absent from populations throughout its distribution (McBride and Jacobs, unpublished data). Exceptions are *Poria sequoia* Fritz & Bonar that causes trunk rot (28) and *Seiridium* sp., a twig and branch canker pathogen reported as a potential killer of seedlings, sprouts and saplings under natural conditions (49). Also, natural regeneration via seedlings is partially prevented in areas covered by redwood litter due to the action of various fungal pathogens that attack seeds and seedlings (12, 21).

The Leaf Fungal Endophytic Community of Redwoods

Leaf endophytes in redwood have been previously reported from trees planted in France (5) and from native trees in northern California and southern Oregon (8). Each report indicated low diversity of endophytes, recording only four species. *Chloroscypha chloromela* (Phil. and Hk.) Seaver was the dominant endophyte in the northernmost portion of redwood distribution, where it was isolated in 94% of the redwood leaves sampled. These endophytes are not systemic and infect the host in winter during the rainy season (15). Thus, we decided to study the endophytic communities of redwood along a geographic transect and in more detail within a single redwood population.

Geographic Survey of Endophytes

A north to south transect representing coastal redwood's geographic range, from southern Oregon to central California, was sampled to determine which endophytes were prevalent in redwood leaves (48). Six sites were sampled, among which were those both never logged and those mature second growth forests (Fig. 1). A minimum of 20 apparently healthy lower canopy leaves per adult tree of age classes one to four years old were processed within 48 hr of collection. Leaves were rigorously surface sterilized with ethanol and sodium hypochlorite solutions, and each leaf was plated onto water agar slants in individual culture tubes.

Infection rate was very high at most sites from 96 to 99%. However, the infection rate at the southernmost site was 81.4%. Rarely, more than one endophyte species was detected from the same leaf (5.4 to 15.0% of leaves had double or multiple species infections). *Pleuroplaconema* sp. was the numerically dominant endophyte across all sites, occurring in 55.4% to 74.6% of the sampled leaves. Typically non-dominant species commonly present in the redwood populations with their respective percent infection of leaves were: *Cryptosporiopsis abietina* Petrak (0-30.0%), *Phomopsis* spp. (0-5.0%), *Nodulisporium* sp. (3.0-15.0%), and *Geniculosporium* sp. (2.0-9.3%). Other species were found regularly but with abundances typically

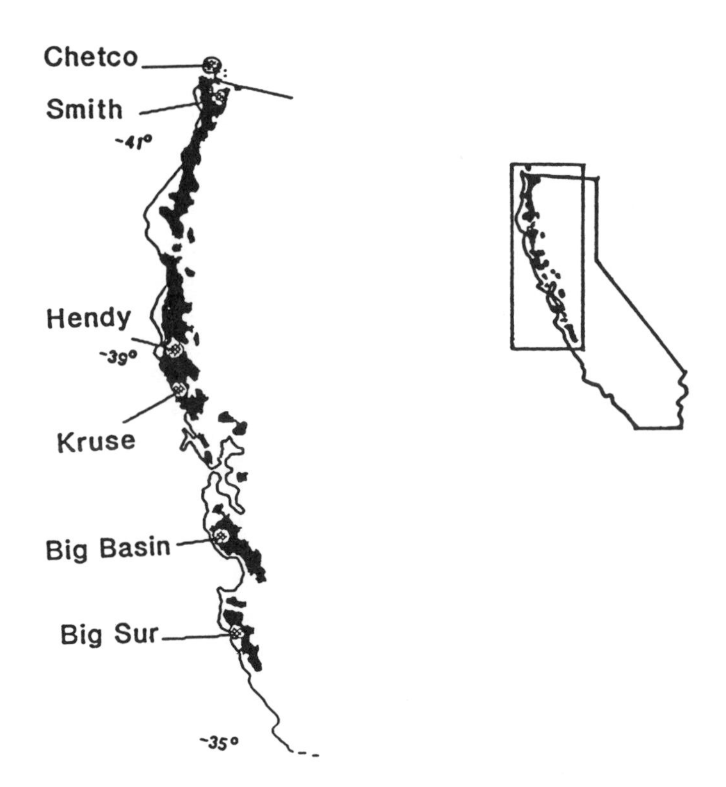

Fig. 1. Location of study sites in central and northern California and southern Oregon for the geographic survey of redwood leaf fungal endophytes in relation to the distribution of coastal redwood. All five California sites were sampled at State Parks, and the Oregon site was taken on National Forest land.

less than five percent.

The presence of endophytes in redwood leaves was relatively constant along the latitudinal transect, and did not represent distinct geographic mycobiotas. Although there were differences in the abundance of some species between sites, the most important species had similar incidences among sites. The most dissimilar site occurred near the northern limit of redwood's natural distribution at Jedediah Smith State Park. This difference was due primarily to the relatively high abundance of one species,

Cryptosporiopsis abietina. A *Pleuroplaconema* sp. was the numerically most important species at all sites.

Detailed Examination of One Central California Site

Although the geographic transect was aimed at general trends, a single redwood population in central California was studied intensively (16). Leaves ranging from one to twelve years old were analyzed in adult trees and basal sprouts to determine endophytic diversity and the possible existence of some spatial patterns in distribution of endophytes. Five leaves per age class per tree or sprout were rigorously surface sterilized and plated onto water agar in petri dishes.

Most leaves (98.7%) were infected by at least one species. Twenty species and six sterile mycelia were detected. *Pleuroplaconema* sp. was the most frequent endophyte (42.6%) followed by *Cryptosporiopsis abietina* (24.4%) [16]. Among the identified species are a suspected mutualist, endophyte generalists, species of unknown ecological status, suspected or actual pathogens, suspected saprobes and a few common inhabitants of leaf phylloplanes. Overall, leaf endophytes in redwood were as common and diverse as in other conifers (8, 42).

Our sampling, unlike that of Carroll and Carroll (8), did not record *Chloroscypha chloromela* at any of the sites. This discrepancy could be explained if *Pleuroplaconema* sp. is an anamorph of *C. chloromela*. This possibility is suggested by our studies, indicating that *Pleuroplaconema* sp. is the most prevalent endophyte in redwood and because it has extremely similar cultural characteristics to *C. chloromela* (O. Petrini, pers. com.). *Pleuroplaconema* sp. might be important for redwood because it meets five out of six criteria proposed by Carroll (6) to recognize likely endophytic mutualists: a) it does not cause apparent symptoms of disease in redwood and there is no report associating it with redwood disease; b) *Pleuroplaconema* sp. is not transmitted by seed (12, Espinosa-García, unpublished results), but is laterally transmitted from tree to tree in a presumably reliable manner (15). All or most trees surveyed to date for endophytes were infected by *Pleuroplaconema* sp. or *C. chloromela* (8, 16, 48). c) although the level of infection by *Pleuroplaconema* sp. within the leaves is not known, it infects a high proportion of the leaves of its host (16, 48) and is present in young as well as old leaves (16); d) *Pleuroplaconema* sp. is ubiquitous throughout the redwood range (48); e) *C. chloromela* shows the same pattern of intracellular infection as *Rhabdocline parkeri* Swd.-Pike, Stone & Carroll, a demonstrated mutualist of Douglas fir (6). Espinosa-García (unpublished results) has also observed this pattern of intracellular infection in redwood leaves from one of our study sites. There is no evidence for or against the sixth criterion, i.e., the production of antibiotics or toxic metabolites by *Pleuroplaconema* sp.

Diversity and equitability in the endophytic community increased rapidly along the leaf age spectrum, until they reached a plateau in leaves three to seven years old. These parameters decreased by almost half in eight to twelve years old leaves. These changes were linked with the statistically significant high frequency of *Cryptosporiopsis abietina* in old leaves and *Pestalotiopsis funerea* (Desm.) Stey. in young leaves. Although the leaf endophytic communities of basal sprouts and mature trees were similar, differences in the distribution of *Pleuroplaconema* sp. and *Pestalotiopsis funerea* were statistically significant, the former being more frequent in trees and the latter in sprouts (16).

Tree age, stress, and different environmental conditions between recently clear-cut and ancient forests may also affect the endophyte species present in redwood leaves. The frequency of *Pleuroplaconema* sp. decreased and that of *Phomopsis* spp. increased in 10 to 15 year old regenerating clear-cut areas relative to ancient forest in the northern part of redwood's distribution (Rollinger, unpublished data). This trend is interesting given that *Phomopsis* spp. have been described as pathogens of redwood seedlings, and that *Pleuroplaconema* sp. is the endophyte most closely associated and most likely to be a mutualist with redwood.

Although some species were differentially distributed according to age, developmental status of the plant, and logging history of the site, the overall pattern of endophytic colonization in leaves of progressing age was patchy without a particular sequence of species succession (48). Fungus infection has been demonstrated in some plants to increase or decrease leaf tissue quality for herbivores or pathogens (6, 29, 30, 36). Therefore, we anticipate that endophytes might differentially modify leaf quality to consumers either by upgrading or degrading this resource, and thus transform branches and whole plants into mosaics of favorable and unfavorable substrates. Variability in plants has been considered an important factor that prevents virulent or devastating consumer strains from becoming widespread (4, 13). Within-plant variability is considered even more important in trees, where one tree generation is exposed to many consumer generations (56). The fact that coastal redwood not only is very long-lived but can reproduce vegetatively results in additional exposure of a single genet to plant consumers. The patchy endophytic distribution in redwood as well as the intrinsic variability of endophytes could provide, in addition to those of redwood secondary chemicals, other sources of variability in leaf tissue quality for redwood herbivores and pathogens.

Interactions Among Readwood Endophytes

During the first stages of our investigations with redwood endophytes, we observed that nearly all leaves older than 1 year were infected with fungal endophytes (16, 48). In addition, the appearance of more than one species per leaf in our cultures was uncommon, even though 26 or more species of fungi occurred endophytically in redwood needles with some of them being quite abundant (16). These observations suggested that perhaps some species either prevented colonization or growth of other species from senesced leaves. Hypothesized mechanisms of this interference were allelopathy or competition for substrate. Due to the great diversity of fungal endophytes in redwood needles and that each of these species might have a distinct effect on potential mutualisms, disease, or leaf decomposition, we decided to explore the interaction among these endophytes. We hypothesized that the presence of certain redwood foliar endophytes affect the growth rate of other redwood endophytes *in vitro* (47).

Bioassays were made with eight endophyte species in which pairs of the species were grown simultaneously at opposite sides of petri dishes on redwood extract agar at both high and low nutrient concentrations. Linear growth was measured at regular time intervals.

Not surprisingly, this diverse group of fungi exhibited different responses growing in the presence of other fungi. In 48 of 78 total species combinations growth was either significantly positively or negatively affected. All species significantly reduced the growth of at least one other species. Growth of pathogenic endophytes, *Botrytis cinerea* (Fr.)Pers., *Pestalotiopsis funerea*, and the possibly pathogenic ones, *Cryptosporiopsis abietina* and *Cucurbitaria* cf. *coronillae* (S.F. Gray) De Thum. was reduced up to 28.5%, 56.7%, 25.1%, and 32.3%, respectively. *Geniculosporium* sp., which has an undetermined role, was stimulated by some fungi and inhibited by others. When grown with other endophytes significant stimulation occurred in the growth of the pathogenic endophyte *Phomopsis occulta* (up to 35.7%) and the possible mutualist, *Pleuroplaconema* sp. (up to 29.5%) [Fig. 2]. Although its ecological role is unclear, *Hypoxylon bipapillatum* Berk. & Curt. both conspicuously inhibited and tended to be stimulated by other fungi.

This evidence suggests that certain redwood endophytes may function as antagonists or stimulators to pathogens, depending on the particular interaction. The non-pathogens that negatively affect pathogens *in vitro* might benefit redwood by competing with or producing chemicals toxic to the pathogens. Stimulators to non-pathogens and possible mutualists may also decrease the impact of pathogens on redwood. The increased fungal biomass could produce more allelochemicals against pathogens or increase the competitive ability of the stimulated fungi.

Overall, the incidence, distribution patterns and potential inter-fungal interactions of possible pathogens, saprobes and an apparent specialist suggest that not only single species but whole endophytic communities may be important for the plants that harbor them.

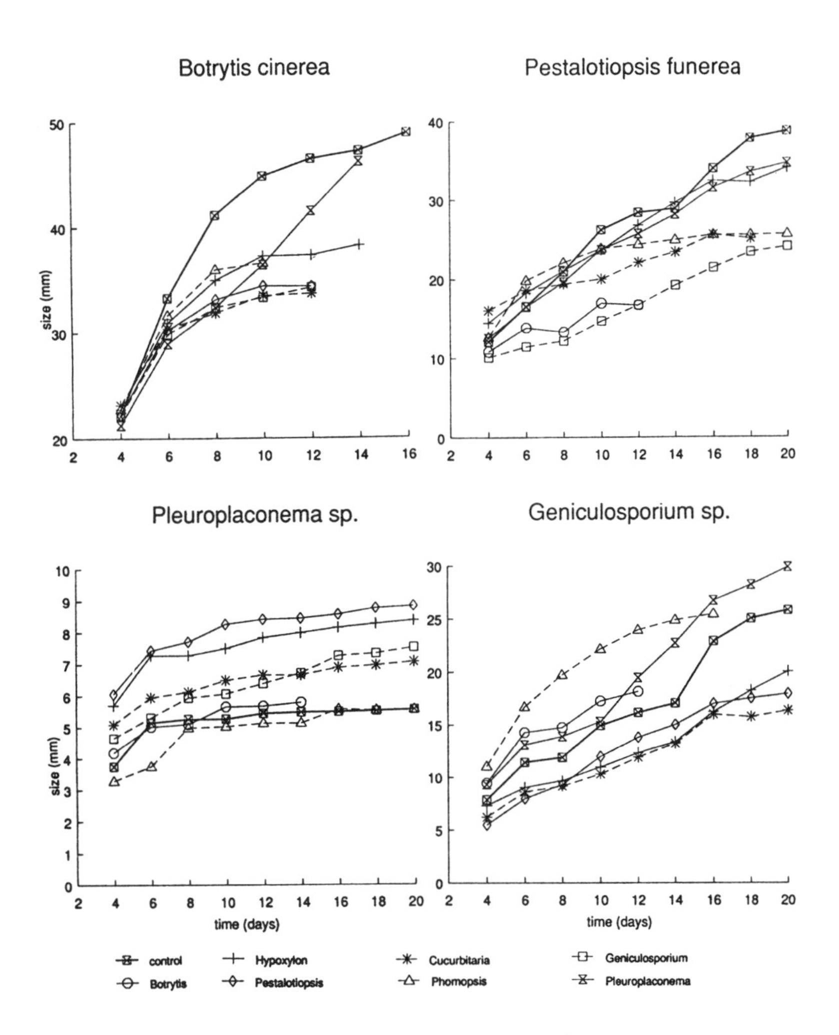

Fig. 2. Example growth curves of two endophytes grown in the presence of other endophytes. *Pleuroplaconema* sp. was grown on the high nutrient concentration medium and *Hypoxylon bipapillatum* on the low nutrient concentration medium.

Leaf Endophyte Response to the Volatile Terpenoids of Their Host

Essential Oils in Coastal Redwood

Because redwood leaf endophytes are ubiquitous and diverse, we decided to study the relationships of these organisms with the volatile terpenoids, which are also an ubiquitous and diverse component of redwood leaves.

Monoterpenoids (C_{10}), and sesquiterpenoids (C_{15}) are very common in plants, and frequently are major components of volatile mixtures, called essential oils. These compounds have antiherbivore and antipathogen activity and are often deemed vital in protecting plants against their consumers (e.g. 10, 33, 37). Volatile terpenoids occur in complex mixtures with repetitive compositional patterns. In trees, the relative proportion of each compound is under tight genetic control and a particular compositional pattern (here called "phenotype") is uniform in the mature foliage of an individual (27, 33, 55). Terpenoid compositional patterns within and among populations of some plants tends to be multimodal, i.e. several terpenoid phenotypes are distributed among many individuals in a population (34, 35, 50). This variability and that of other secondary chemicals are considered important factors preventing pathogen epidemics and herbivore outbreaks (43).

Redwood leaves have relatively large amounts of essential oils accumulated in resin ducts, typically two marginal and one along the midrib (11). Monoterpenes constitute 60-88% of the essential oil and account for 0.1 to 6.6% of the leaf dry weight (23). The other essential oil components are sesquiterpenes, oxygenated terpenoids and, in trace amounts, allyl phenylethers (22, 40).

Redwood leaf essential oil varies both within the population and developmentally within the individual, both in yield (i.e. total amount) as well as in compositional pattern (i.e. percentage of individual compounds relative to the total amount) (23). The monoterpene yield and compositional pattern (i.e. monoterpene phenotype) of a leaf changes during its first six months of life and then stabilizes until senescence starts. The monoterpene phenotype is constant in the mature leaves of a tree, but the yield of the uppermost canopy leaves is higher than that of the other leaves (24, 25, 26). Leaf monoterpene yield in basal sprouts may be two- or three-fold higher than in leaves of the same age from the parent tree; the monoterpene phenotype may also change, with sabinene and γ-terpinene increasing in relative importance (24)

Essential oil phenotypes in mature redwood leaves have been recognized in a central California population (16). These phenotypes were characterized by the spectrum of five predominant monoterpenes (sabinene, limonene, β-phellandrene, γ-terpinene and α-pinene) and the sesquiterpene caryophyllene, although 9 additional monoterpenes were also analyzed. Four redwood phenotypes were selected to explore their effect on endophytes. In three of the Phenotypes (I,II and IV), sabinene is the most abundant compound, followed by relatively similar but characteristically different amounts of limonene, γ-terpinene, β-phellandrene and α-pinene. In Phenotype III, however, β-phellandrene occurs in the highest amount with α-pinene, sabinene and limonene being relatively similar in concentration. The relatively high amount of caryophyllene also distinguishes Phenotype I.

Endophytes and Essential Oil Phenotypes

Seven of the potentially important coastal redwood endophytes were selected to explore the potential role of four redwood leaf essential oil phenotypes in controlling endophytic activity. The endophyte species selected included a suspected mutualist, pathogens and a species never found in redwood. Effects on endophytic growth *in vitro* of the phenotype of essential oil, its concentration and two of its components were assessed.

The response of 14 isolates of the suspected mutualist *Pleuroplaconema* sp. to essential oil phenotypes was studied first (17). Two hypotheses were tested: a) distinctive patterns of leaf essential oils extracted from four trees would have a differential effect *in vitro* on the growth of *Pleuroplaconema* sp., and b) the growth of fungal isolates from a particular tree would be differentially more negatively affected when exposed to essential oils phenotypes different from that of their tree of origin. Each isolate was exposed to each of the essential oil phenotypes in chambers with atmospheres initially saturated with these chemicals. Linear growth of the isolates was measured after exposure.

The essential oil phenotypes did inhibit the growth of the endophyte differentially, but the inhibition pattern did not support the second hypothesis. The fungal isolates from a particular tree did not show strong differential reaction to the essential oil of their tree of origin. *Pleuroplaconema* sp., the predominant redwood endophyte, displayed low average tolerance to the essential oils of its host as well as a reduced variability in susceptibility when the individual isolates were considered. These responses were opposite to those of conifer specialized pathogens, such as *Crumenulopsis sororia*, a pine canker pathogen that exhibited high tolerance to its host monoterpenes with wide variability in susceptibility to some of these compounds (14). These unexpected results led to the hypothesis that essential oils are an important factor controlling the activity of *Pleuroplaconema* sp. after it colonizes the leaf. Moreover, we speculated that the high susceptibility of this endophyte to its host terpenoids was the outcome of a process in which decrease in pathogenicity is concurrent with increased susceptibility to the secondary compounds of the host.

The sensitivity of *Pleuroplaconema* sp. to its host terpenoids may be explained if these compounds are important in maintaining the apparent inactivity of the fungus after colonization. The fungus would become active following a decrease in concentration of essential oils in the leaf, as it occurs during redwood leaf senescence (25). (The activation of the fungus in a non-senescent leave might occur with a physiological alteration preventing the replenishment of terpenoids of the leaf. We speculate that such alteration could be herbivore attack, e.g. gall forming insects.) A preformed antifungal compound in the avocado fruit is the cause of the latency of the pathogen *Colletotrichum gloeosporioides* (45), which becomes active when the antifungal compound is degraded during fruit ripening (46). Krupa and Fries (31) demonstrated that *Boletus variegatus*, an ectomycorrhizal fungus, induces high amounts of volatile terpenoids when it colonizes roots of *Pinus sylvestris*. This induction, as well as the antifungus properties of terpenoids led these authors to propose a model in which host terpenoids, and other non-volatile secondary compounds would restrict the growth of the fungus symbiont and maintain the mutualistic state. The redwood-*Pleuroplaconema* sp. relationship might fit a similar model, but with preformed terpenoids as the controlling agents.

Further exploration of the effect of leaf essential oil phenotypes on endophytic growth was pursued with the other six selected species (18). We tested the hypothesis that leaf essential oil phenotypes would have differential effects on endophytic species. We analyzed experimentally the overall effects *in vitro* of four redwood essential oil phenotypes on six fungal species. Additionally, we considered the differences in response to redwood essential oils of the following: 1) a generalist pathogen that attacks redwood [*Botrytis cinerea*]; 2) three species of actual or potential conifer pathogens [*Pestalotiopsis funerea*, *Phomopsis occulta* Trav. and *Seiridium juniperi* (All.) Sutton], isolated as leaf endophytes in redwood; 3) a common conifer endophyte of uncertain ecological status [*Cryptosporiopsis abietina*], isolated from redwood; 4) and a mutualistic endophyte [*Meria* state of *Rhabdocline parkeri*] known only from Douglas Fir [*Pseudotsuga menziesii* (Mirb.) Franco]. Intraspecific variability of fungi in response to terpenoids can be significant (14, 52). Thus, we further analyzed five isolates of each *C. abietina*, *P. funerea* and *P. occulta*. Unfortunately only one isolate for each of the remaining species was available at the time of the experiment. Thus, conclusions from this experiment, particularly those for one-isolate species, should be taken carefully. Isolates were exposed to essential oils after adjusting the exposure time to the growth rate of the endophyte species and measuring linear growth afterwards.

The four essential oil phenotypes were uniformly inhibitory for some endophyte species and differentially so for others (Fig. 3). Susceptibility varied widely among fungal species to the volatiles of the four phenotypes. Within a species, the largest differences among isolates occurred in the overall tolerance to the phenotypes. However, the order of inhibition intensity among isolates to the phenotypes was very similar in *C. abietina* and *P. funerea*. Isolates of *P. occulta* were more variable in response, but the differences in reaction to the redwood phenotypes were small. The more contrasting response pattern to the phenotypes occurred in *S. juniperi*,

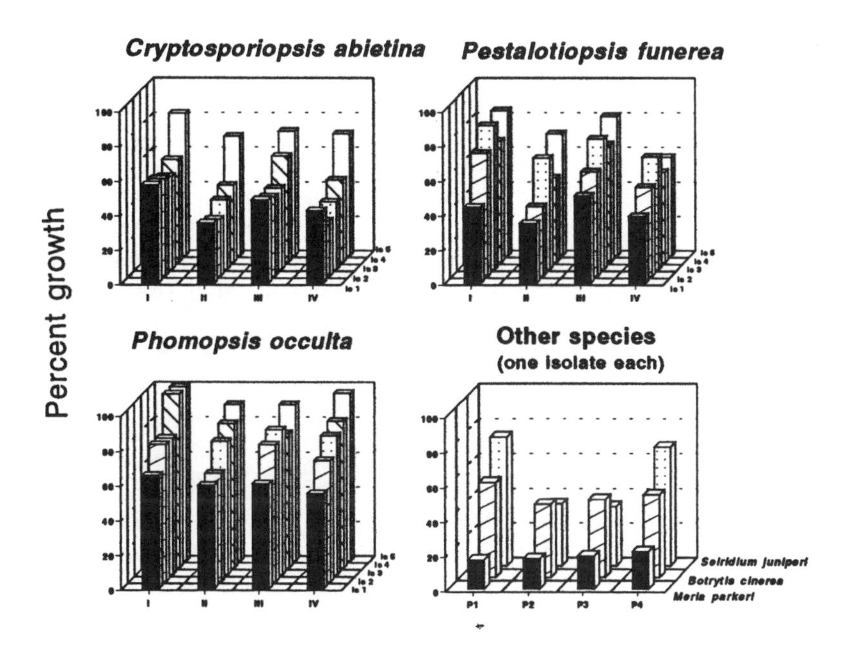

Fig. 3. Growth response of isolates of six endophytic species to four essential oil phenotypes from coastal redwood. Is=Isolate. Bars represent per cent growth relative to control. For *Cryptosporiopsis abietina*, *Pestalotiopsis funerea* and *Phomopsis occulta*, each row represents the response of one isolate. *Sieridium juniperi*, *Botrytis cinerea* and *Meria parkeri* were represented by one isolate each. Data from 18, used by permission.

the more pathogenic fungus of the group. The examination of more *S. junipera* isolates is needed to determine if this response pattern is repeated among individuals of this species.

The specialized pathogens were the least susceptible to the essential oils and the Douglas fir endophyte was the most; *B. cinerea* and *C. abietina* displayed intermediate susceptibility. When *Pleuroplaconema* sp. is compared to this group of endophytes, the suspected mutualist shows a higher susceptibility than the conifer specialized pathogens to these terpenoids (17). Additionally the overall tolerance of *B. cinerea* and *C. abietina* was very similar to *Pleuroplaconema* sp., but still they were less inhibited than the latter fungus. Only *Meria parkeri* from Douglas fir was more susceptible to these terpenoids. Experimentation with more isolates from all species would be desirable to confirm these results.

Low virulence in some pathogenic fungi has been associated with inability to break down host secondary compounds formed after fungus infection, i.e. phytoalexins (53). Although the phytoalexins in those pathosystems were phenolic compounds, a similar process might occur with either preformed terpenoids or those that are induced. The higher sensitivity of *Pleuroplaconema* sp. to redwood terpenoids than the possible pathogenic endophytes may be associated with a process of increased susceptibility to the host secondary chemicals with concurrent pathogenicity attenuation. A possible relationship between increased specialization in endophytic lifestyle, with low or null pathogenic potential and high susceptibility to the host secondary chemicals, may be worth exploring in plant-endophyte systems where mutualism is known or suspected. The sensitivity of some pine ectomycorrhizae and root pathogens to the host's terpenoids is consistent with this idea. A study in a pine-fungi system revealed that some of the induced terpenoids were much more inhibitory to the ectomycorrhizal fungus than to a root pathogenic fungus (44, 45).

Our results indicate that relatively similar but distinctive redwood essential oils occurring naturally had a differential activity on endophytes. This suggests that essential oil variability is important in redwood-endophytes interactions, and that the diversity of responses shown by endophytic species to essential oils may have a differential intra- and interspecific importance in preventing endophytic growth within redwood foliage.

Endophytes and Essential Oil Dosage

The essential oil phenotype can determine the degree of inhibition for some endophytes. However, dosage, as well as composition, might be important in contributing to control of endophytic activity. Because significant variability in leaf essential oil yield occurs among redwood trees (23), and decrease in total amount of these terpenoids in senescent leaves has been documented (38), the dose-dependent response to essential oils of *Pleuroplaconema* sp. and *Pestalotiopsis funerea* growing *in vitro* on culture media with different concentration of nutrients was studied (20). We designed an experiment to test the hypothesis that essential oils are a causal factor in the inactivity of highly specialized endophytes. Four isolates of the redwood specialized endophyte *Pleuroplaconema* sp., and four of *Pestalotiopsis funerea*, a conifer-generalist endophyte, were exposed *in vitro* to increasing essential oil doses. The latter species was included for comparison. This exposure was performed with the fungi growing either in a standard redwood medium ("full strength") or a 50% dilution of this medium with water ("half strength"). We attempted in this manner to roughly simulate the nutrient quality for endophytes of a mature leaf, where nutrients for endophytes are scarce (half strength medium) and that from a senescent leaf, where nutrients are released and available to endophytes (full strength medium).

Pleuroplaconema sp. was stimulated with essential oil dosages, and then it was increasingly inhibited, with essentially no growth in a saturated atmosphere of terpenoids (Fig. 4A). By contrast, *Pestalotiopsis funerea* was unaffected by low volatile concentrations and inhibited by medium and high concentrations (Fig. 4B). Moreover, *P. funerea* was not inhibited more than 80% even in the two highest essential oil dosages. Statistically significant differences were not found in the response of *Pleuroplaconema* sp. to dilution of the culture medium. *P. funerea* did respond to the dilution of the culture medium, but differences were small and can be attributed to the response of only two isolates. Although the effect of nutritional factors on endophyte latency needs to be explored further, the overwhelming effect of essential oil dosage in either strength of culture medium reinforces the idea that these chemicals are a major factor controlling endophytic activity. Our study may be similar to that of *Colletotrichum gloeosporioides* where nutrients were discarded as factors affecting fungus latency in avocado fruits, and the concentration of a secondary chemical in the avocado peel was found to determine of the activity or latency of this pathogen (44, 45, 46).

Our results support the hypothesis that essential oils, at least partially, control the activity of *Pleuroplaconema* sp. Redwood essential oils may not completely suppress the activity of *Pestalotiopsis funerea*, but they could be very important in conjunction with other redwood secondary chemicals, such as diterpene acids, bisflavones and proanthocyanidins (38, 39, 51) or stress factors that can increase fungal susceptibility to essential oils (32) and other secondary chemicals. Thus, stress agents in leaves along with essential oils and other secondary chemicals may control the activity of some endophytes.

Endophyte Response to Components of Leaf Essential Oils

The differential effects of these distinctive leaf essential oil phenotypes on endophytes, in spite of their similarities in chemical profile, indicated the possibility that their components were acting in a synergistic or additive fashion. Two important components of redwood essential oil, sabinene and γ-terpinene, were used to explore this possibility *in vitro* (19). Both type and dose of essential oils proved to be potentially important in contributing

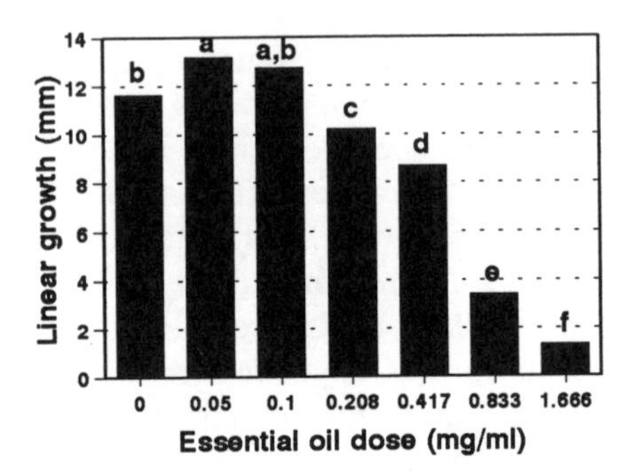

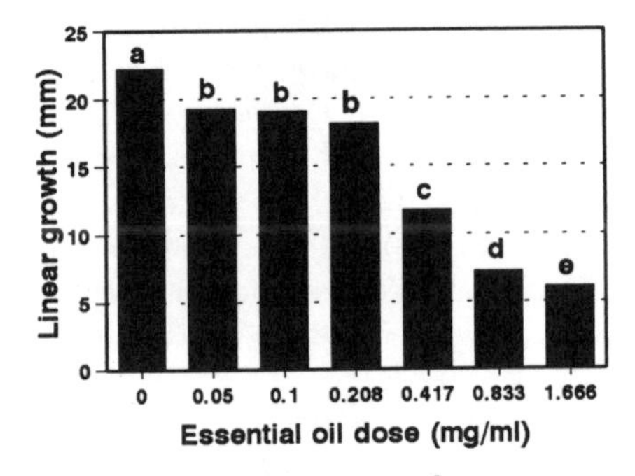

Fig. 4. Overall dose response of A) *Pleuroplaconema* sp. and B) *Pestalotiopsis funerea* to redwood leaf essential oils. Each bar represents an average of 48 measurements. For each species, bars with the same letter are not significantly different at alpha=0.05 after Duncan's Multiple Range Test. Data from 20, used by permission.

to reduction of endophytic growth. Thus, mixtures with different ratios of these compounds, as well as different concentrations of these compounds acting alone, were assayed on six endophytic species.

Sabinene and γ-terpinene, alone and in mixtures, have inhibitory effects on the species tested. Dose-dependent response to these monoterpenes in single compound trials varied according to species. With low concentrations of these individual monoterpenes (0.375 to 1.5 mg/6 ml air), most species reduced their growth from 10 to 60% of the control. However, saturated atmospheres of sabinene or γ-terpinene essentially stopped the growth of most species. *Pestalotiopsis funerea* growth was inhibited about 80 % with saturated atmospheres of both monoterpenes, a response similar to that elicited by high doses of leaf essential oil. *Seiridium juniperi* was completely inhibited by a saturated atmosphere of γ-terpinene, but that of sabinene never decreased the growth of the fungus more than 50%.

Effects of mixtures of different ratios suggest that sabinene and γ-terpinene act in an additive fashion, inhibiting the growth of *Botrytis cinerea*, *Cryptosporiopsis abietina*, *Pleuroplaconema* sp., *Phomopsis occulta*, *Pestalotiopsis funerea* and *Seiridium juniperi* (Table 1).

The synergistic and additive effects observed for these monoterpenes, as well as those that may be found with other components of the essential oils, are likely to enhance the inhibitory effects of essential oils. They can make the natural selection of fungi able to deal with such a diverse chemical mixture more difficult, as essential oils (both yield and compositional pattern) are also not uniform throughout the host population.

Variability and diversity were the hallmarks in this research. Although the redwood leaf endophyte communities were similar across the redwood distribution, leaf endophytes were also diverse in coastal redwood. They had a variable pattern of colonization and showed a diversity of interactions among themselves. Some of these endophytes are likely to be latent pathogens, dormant saprobes, and *Pleuroplaconema* sp. (the numerically dominant species throughout redwood's distribution) may be a mutualist with redwood. Redwood has distinct leaf essential oil phenotypes characterized by a diverse array of chemical components. These essential oils had varied effects on endophytes dependent on essential oil type and dose, both of which are variable in redwood populations. Finally, endophytic response was variable within and among species, and that response might be linked with the diverse ecological status of endophytes.

Table 1. Endophytic growth response (mm) to sabinene, γ-terpinene and mixtures of them. In all cases 1.5 mg per vial of compound or mixture were added. Expected growth values for concentrations of 1.5 mg of single compounds or mixtures per vial are in parentheses. Values were calculated with regression equations and Wadley's formula. Expected and actual growth under mixtures were compared with X^2 tests and significant differences were not detected at $p = 0.05$.

	Sabinene	Mixtures			γ-Terpinene	X^2
	100-0	25:75	50:50	75:25	0-100	
B. c.	11.0	16.5	16.0	17.5	16.5	
	(14.7)	(14.8)	(16.0)	(14.36)	(15.0)	1.03
C. a.	5.3	-.-[1]	4.8	5.3	4.3	
	(4.7)	(4.7)	(4.7)	(4.5)	0.072	
P. f.	16.3	13.8	13.5	15.3	14.2	
	(13.2)	(13.1)	(13.1)	(13.2)	(13.0)	0.38
P. o.	9.8	10.0	8.9	8.5	9.0	
	(9.1)	(8.8)	(8.9)	(9.0)	(8.7)	0.18
P. sp.	3.0	4.6	5.0	3.3	3.3	
	(3.1)	(3.5)	(3.4)	(3.2)	3.7)	1.13
S. j.	5.8	3.8	4.6	4.6	5.8	
	(5.4)	(4.3)	(4.8)	(5.0)	(4.1)	0.09

B.c.= *Botrytis cinerea; C.a.*= *Cryptosporiopsis abietina; P.f.*= *Pestalotiopsis funerea; P.o.*= *Phomopsis occulta; P.* sp.= *Pleuroplaconema* sp.; *S.j.*= *Seiridium juniperi*

[1] Missing data. The effect of this mixture was assessed in a further separate experiment. Means were: sabinene, 7.1 mm; 25:75 mixture, 6.0 mm; γ-terpinene, 6.7 mm. Differences among means are not statistically significant after one-way ANOVA. Data from 19, used by permission.

ACKNOWLEDGEMENTS

We are grateful to Dr. G. Carroll, Dr. F. Cobb, Dr. V. Jaramillo and Dr. O. Petrini for their comments and suggestions. This work was partially funded by U.C.S.C. Faculty Research funds grant to J.H.L., and graduate student grants to F.J.E. and J.R. from the U.C.S.C. Biology Department. F.J.E. wishes to thank the Universidad Nacional Autónoma de México and México's Consejo Nacional de Ciencia y Tecnología for supporting him with a fellowship during the period of this study.

LITERATURE CITED

1. Bacon, C.W., Lyons, P.C., Porter, J.K. and Robbins, J.D. 1986. Ergot toxicity from endophyte-infected grasses: a review. Agron. J. 78: 106-116.
2. Bose, S.R. 1956. *Alternaria* within the pericarp of the wheat seed. Nature 178: 640-641.
3. Bradshaw, A.D. 1959. Population differentiation in *Agrostis tenuis* Sibth. II. The incidence and significance of infection by *Epichloë typhina*. New Phytol. 58: 310-315.
4. Burdon, J.J. and Chilvers, G.A. 1982. Host density as a factor in plant disease ecology. Annu. Rev. Phytopath., 20: 143-166.
5. Carroll, F.E., Müller, E. and Sutton, B.C. 1977. Preliminary studies on the incidence of needle endophytes in some European conifers. Sydowia 29: 87-103.
6. Carroll, G.C. 1986. The biology of endophytism in plants with particular reference to woody plants. Pages 205-222 in: Microbiology of the Phyllosphere. N. J. Fokkema and J. van den Heuvel, eds. Cambridge University Press, Cambridge.
7. Carroll, G.C. 1988. Fungal endophytes in stems and leaves: from latent pathogen to mutualistic symbiont. Ecology 69: 2-9.
8. Carroll, G.C. and Carroll, F.E. 1978. Studies on the incidence of coniferous needle endophytes in the Pacific Northwest. Can. J. Bot. 56: 3034-3043.
9. Chapela, I.H. and Boddy, L. 1988. The fate of early colonizers in beech branches decomposing on the forest floor. FEMS Microbiol. Ecol. 53: 273-284.
10. Cobb, F.W. Jr., Krstic, M., Zavarin E. and Barber H.W. 1968. Inhibitory effects of volatile oleoresin components on *Fomes annosus* and four *Ceratocystis* species. Phytopathology 58: 1327-1335.
11. Dallimore, W. and Jackson, A.B. 1966. A Handbook of the Coniferae and Ginkoaceae. St. Martin's Press, New York. 729 pp.
12. Davidson, J.G.N. 1971. Pathological Problems in Redwood Regeneration from Seed. Ph.D. Dissertation. University of California, Berkeley. 288 pp.

13. Denno, R.F. and McClure, M.S. 1983. Variable Plants and Herbivores in Natural and Managed Systems. Academic Press. New York. 717 pp.
14. Ennos, R.A. and Swales, K.W. 1988. Genetic variation in tolerance to host monoterpenes in a population of the ascomycete canker pathogen *Crumenulopsis sororia*. Plant Pathol. 37: 407-416.
15. Espinosa-García, F.J. 1991. Studies of the Relation of the Fungal Endophytic Community and Essential Oils in the Leaves of Coastal Redwood (*Sequoia sempervirens*). Ph.D. Dissertation, University of California, Santa Cruz. 147 pp.
16. Espinosa-García, F.J. and Langenheim, J.H. 1990. The endophytic fungal community in leaves of a coastal redwood population. Diversity and spatial patterns. New Phytol. 116: 89-97.
17. Espinosa-García, F.J. and Langenheim, J.H. 1991a. Effect of some leaf essential oil phenotypes from coastal redwood on the growth of its predominant endophytic fungus, *Pleuroplaconema* sp. J. Chem. Ecol. 17: 1837-1857.
18. Espinosa-García, F.J. and Langenheim, J.H. 1991b. Effect of some essential oil phenotypes in Coastal Redwood on the growth of several fungi with endophytic stages. Biochem. Syst. Ecol. 19: 629-642.
19. Espinosa-García, F.J. and Langenheim, J.H. 1991c. Effects of sabinene and γ-terpinene from coastal redwood leaves acting singly or in mixtures on the growth of some of their fungus endophytes. Biochem. Syst. Ecol. 19: 643-650.
20. Espinosa-García, F.J., Saldívar-García, P. and Langenheim, J.H. 1993. Dose-dependent effects *in vitro* of essential oils on the growth of two endophytic fungi in coastal redwood leaves. Biochem. Syst. Ecol. 21: 185-194.
21. Florence, R.G. 1965. Decline of old-growth redwood forests in relation to some soil microbial processes. Ecology 46: 52-64.
22. Gregonis, D.E., Portwood, R.D., Davidson, W.H., Durfee, D.A. and Levinson, A.S. 1968. Volatile oils from foliage of coast redwood and big tree. Phytochemistry 7: 975-981.
23. Hall, G.D. 1985. Leaf Monoterpenes of Coast Redwood (*Sequoia sempervirens*). Ph.D. Dissertation. University of California, Santa Cruz. 150 pp.
24. Hall, G.D. and Langenheim, J.H. 1986a. Within-tree spatial variation in the leaf monoterpenes of *Sequoia sempervirens*. Biochem. Syst. Ecol. 14: 625-632.
25. Hall, G.D. and Langenheim, J.H. 1986b. Temporal changes in the leaf monoterpenes of *Sequoia sempervirens*. Biochem. Syst. Ecol. 14: 61-69.
26. Hall, G.D. and Langenheim, J.H. 1987. Geographic variation in leaf monoterpenes of *Sequoia sempervirens*. Biochem. Syst. Ecol. 15: 31-43.
27. Hanover, J.W. 1966. Environmental variation in the monoterpenes of *Pinus monticola* Dougl. Phytochemistry 5: 713-717.

28. Hepting, G.H. 1971. Diseases of Forest and Shade Trees of the United States. U.S. Dept. of Agric. Handbook No. 360. Washington, DC. 658 pp.
29. Kingsley, P., Scriber, J.M., Grau, C.R. and Delwiche, P.A. 1983. Feeding and growth performance of *Spodoptera eridana* (Noctuidae Lepidoptera) on "vernal" alfalfa, as influenced by *Verticillium* wilt. Prot. Ecol. 5: 127-134.
30. Kirfman, G.W., Branderburg, R.L. and Garner, G.B. 1986. Relationship between insect abundance and endophyte infestation level in tall fescue in Missouri. J. Kansas Entomological Soc. 59: 552-554.
31. Krupa, S. and Fries, N. 1971. Studies on ectomycorrhizae of pine. I. Production of volatile organic compounds. Can. J. Bot. 49: 1425-1431.
32. Kurita, N. and Koike, S. 1982. Synergistic antimicrobial effect of sodium chloride and essential oil components. Agric. Biol. Chem. 46: 159-165.
33. Langenheim, J.H. 1984. Role of plant secondary compounds in wet tropical ecosystems. Pages 189-208 in: Physiological Ecology of Plants in the Wet Tropics. E. Medina, H. Mooney and C. Vasquez-Yanes, eds. W. Junk, The Hague.
34. Langenheim, J.H. and Stubblebine, W.H. 1983. Variation in leaf resin composition between parent tree and progeny in *Hymenaea*: implications for herbivory in the humid tropics. Biochem. Syst. Ecol. 11: 97-106. 31
35. Leather, S.R., Watt, A.D. and Forrest, G.I. 1987. Insect induced chemical changes in young lodgepole pine (*Pinus contorta*): the effect of previous defoliation on oviposition, growth and survival of the pine beauty moth, *Panolis flammea*. Ecol. Entomol. 12: 275-281.
36. Matta, A. 1971. Microbial penetration and immunization of uncongenital host plants. Ann. Rev. Phytopath. 9: 387-410.
37. Miller, R.H., Berryman, A.A. and Ryan, C.A. 1986. Biotic elicitors of defense reactions in lodgepole pine. Phytochemistry 25: 611-612.
38. Miura, H. and Kawano, N. 1968. Sequoiaflavone in the leaves of *Sequoia sempervirens* and *Cunninghamia lanceolata* var. *Konishii* and its formation by partial demethylation. Yakugaku Zasshi 88: 1489-1491.
39. Ohta, K. and Nawamiki, T. 1978. (+)-Polyalthic acid, a repellent against a sea snail *Monodonta neritoides*. Agric. Biol. Chem. 42: 1957-1958.
40. Okamoto, R.A., B.O. Ellison and R.E. Kepner. 1981. Volatile terpenes in *Sequoia sempervirens* foliage. Changes in composition during maturation. J. Agric. Food Chem. 29: 324-326.
41. Petrini, O. 1986. Taxonomy of endophytic fungi of aerial plant tissues. Pages 175-187 in: Microbiology of the Phyllosphere. N. J. Fokkema and J. van den Heuvel, eds. Cambridge University Press, Cambridge.
42. Petrini, O. and Carroll, G. 1981. Endophytic fungi in foliage of some Cupressaceae in Oregon. Can. J. Bot. 59: 629-636.
43. Pimentel, D. and Belotti, A.C. 1976. Parasite-host population and genetic stability. Amer. Natur. 110: 877-888.

44. Prusky, D., Ascarelli, A. and Jacoby, B. 1984. Lack of involvement of nutrients in the latency of *Colletotrichum gloeosporioides* in unripe avocado fruits. Phytopath. Z. 110: 106-109.
45. Prusky, D., Keen, T. and Eaks, I. 1983. Further evidence for the involvement of a preformed antifungal compound in latency of *Colletotrichum gloeosporioides* in unripe avocado fruits. Physiol. Plant Path. 22: 189-198.
46. Prusky, D., Kobiler, I., Jacoby, B., Sims, J.J. and Midland, S.L. 1985. Inhibitors of avocado lipoxygenase: their possible relationship with the latency of *Colletotrichum gloesporioides*. Physiol. Plant Path. 27: 269-279.
47. Rollinger, J., Espinosa-García, F.J. and Langenheim, J.H. 1992. Interaction of redwood endophytes that influences their growth. Bulletin of the Ecological Society of America 17: 306.
48. Rollinger, J. and Langenheim, J.H. 1992. Geographic survey in fungal endophyte community composition in leaves of redwood. Mycologia 85: 149-156.
49. Roy, D.F. 1966. Sylvical Characteristics of Redwood [*Sequoia sempervirens* (D. Don.) Endl.]. U.S.D.A. Forest Service. Research Paper PSW-28. 20 pp.
50. Squillance, A.H., Powers, Jr. H.R. and Kossuth, S.V. 1985. Monoterpene phenotypes in loblolly pine populations: natural selection trends and implications. Pages 299-308 in: Proc. 18th South. For. Tree Improve. Conf.
51. Stafford, H.A. and Lester, H.H. 1986. Proanthocyanidins in needles from six genera of the Taxodiaceae. Am. J. Bot. 73: 1555-1562.
52. Thibault-Balesdent, M. and Delatour, C. 1985. Variabilite du comportament de *Heterobasidion annosum* (Fr.) Bref. a trois monoterpenes. Eur. J. For. Path. 15: 301-307.
53. VanEtten, H.D., Matthews, D.E. and Matthews, P.S. 1989. Phytoalexin detoxification: importance for pathogenicity and practical implications. Annu. Rev. Phytopath.. 27: 143-164.
54. Verhoeff, K. 1974. Latent infections by fungi. Annu. Rev. Phytopathol. 12: 99-110.
55. Von Rudloff, E. 1975. Volatile leaf oil analysis in chemosystematic studies of North American conifers. Biochem. Syst. Ecol. 2: 131-167.
56. Whitham, T.G. (1983). Host manipulation of parasites: within-plant variation as a defense against rapidly evolving pest. Pages 15-42 in: Variable Plants and Herbivores in Natural and Managed Systems. R.F. Denno and M.S. McClure, eds. Academic Press. New York .

CHAPTER 6

FUNGAL ENDOPHYTES OF PALMS

Katia Ferreira Rodrigues

The New York Botanical Garden
Bronx, NY, 10458-5126,

All living plants so far investigated have been shown to harbor fungi inside their tissues. Such fungi colonizing the inner part of aerial plant tissues have been referred to as "endophytes" and are now known to be widespread in nature (29). This term, as originally used by De Bary (15) referred to any organism occurring within plant tissues, distinct from the epiphytes that live on plant surfaces. Carroll's (8) definition describes fungi that colonize aerial parts of living plant tissues and do not induce visible signs of disease. Pathogenic fungi and mycorrhizae are excluded from his concept. Petrini (30) proposes an expansion of Carroll's definition and incorporates into his concept latent pathogens known to live symptomlessly inside the host tissues during part of their life cycle. In this manuscript the term endophyte is used in a broad sense to include those fungi that spend all or nearly all of their life cycle in the host plant tissue (*sensu* De Bary).

Fungal endophytes of grasses (Poaceae) and sedges (Cyperaceae and Juncaceae) are probably the most extensively studied group (see 12, 13, 43). Reports on the presence of endophytes in vascular plants, other than grasses, have focused mainly on ericaceous, dicotyledoneous plants and conifers (29). A number of factors, i. e., seasonal changes, collection site, and age of the host plant and foliage have been reported to influence the species composition and frequency of the endophyte assemblages (30).

Many papers have documented the presence of internal fungi causing asymptomatic infection in the aerial organs of a vast array of evergreen and deciduous phanerogams. Only recently, however, the presence of fungi living endophytically in the leaf tissues of Palmae has been demonstrated. A preliminary account of endophytes in the Australian fan palm *Licuala ramsayi* (Muell.) Domin. (37), and a more intensive work with a palm from the Amazonian rainforests of Brazil, *Euterpe oleracea* Mart.(36) comprise the only investigations on this host plant family known to date.

This chapter summarizes the information on leaf endophytes in Palmae, with particular emphasis placed on the endophytic communities in *E. oleracea.*

Endophytes of Tropical Plants

General considerations

Most of the investigations on endophytes have been carried out in the Northern hemisphere (see 5, 29, 30) and subtropical regions such as Argentina (2, 6, 18) and New Zealand (24, 25, 33). There are few reports of endophytes from plant material collected in tropical regions. Endophyte-infected plants reported from South America include Araceae, Bromeliaceae and Orchidaceae from French Guyana (32), and from Brazil and Colombia (16), *Eucalyptus viminalis* Labill. (2, 18) and *Baccharis coridifolia* DC. (3, 6) from Argentina, *Euterpe oleracea* (35, 36, 38, 39) from Brazil, and *Trichachne insularis* (L.) Nees (49) from Argentina, Brazil, and Chile.

Unlike the much studied relationships between the Clavicipitaceae and Poaceae, the effect of the fungus/host interrelationships among most other plants is not known. While insect aversion to endophyte infected grasses (14) and conifer (9, 42) leaves is known, there have been no similar observations made in tropical settings. Similarly, there are no observations on the modification of plant growth or habit in tropical plants such as have been made by Latch and collaborators (26) in grasses. Endophytes have been suspected to be the cause of a toxic syndrome in livestock feeding on shrubby species of *Baccharis* (Asteraceae) in Brazil (22), and trichothecene toxins have been produced by endophytic strains of a *Phomopsis* species and by *Ceratopycnidium bacchariicicola* Bertoni & Cabral that were isolated from *B. coridifolia* (48).

With the greatest diversity of plants being found in tropical regions (1), it is evident that the study of tropical endophytes is at a very early stage. The following discussion will be based mainly on data from investigations undertaken with *Euterpe oleracea*, an economically important palm in South America, that is most common in floodplains of the Amazon Basin. Leaf samples of *E. oleracea* belonging to three age-classes [unopened (L1), newly expanded (L2), and mature leaf (L3)] of ten 6- to 18-month-old saplings, and of ten 5- to 20-year-old trees were taken over a two year period from plots located at two subsites designated low várzea (elevation < 1 m), and high várzea (elevation > 1 m). At each sampling period leaves were processed according to Rodrigues (38). The study site is a periodically inundated area known as the Amazonian floodplains or "várzea" (23, 46). The forest in the low várzea is dominated by highly flood-tolerant tree species such as *E. oleracea*, and to a lesser extent by other palms such as *Mauritia flexuosa* L., *Raphia taedigera* Mart., and a monocotyledoneous shrub *Montrichardia arborescens* Schott. The forest in the high várzea is characterized by a higher diversity of species. The forest communities are represented by a large number of dicotyledoneous trees belonging to the Anacardiaceae, Burseraceae, Euphorbiaceae, Fabaceae, Guttiferae, and Myristicaceae. *E. oleracea* is found also in the high várzea, but it is somewhat scarce.

Endophytes of the fan palm, *Licuala ramsayi*, were isolated from leaves of a 10-year-old tree located in a tropical lowland rainforest of Australia (Queensland) during September 1988. The quantitative and qualitative composition of endophytic fungi from the fan palm leaves was determined as previously described (37).

Endophytic Communities of Palm Leaves

Overall colonization

Euterpe oleracea presented a low overall endophyte colonization of the leaves sampled (mean of 25%, during four sampling periods). Rodrigues and Samuels (37), in their investigation of endophytes of *L. ramsayi*, also found low overall colonization rates. This low proportion of fungal colonization is similar to those reported by Carroll and Carroll (10) for the conifers *Abies amabilis* (Dougl.) Forbes (20.4%), *A. lasiocarpa* (Hook.) Nutt. (21.1%), and *Taxus brevifolia* Nutt. (24.6%) in western Oregon and southwestern Washington. The low rates of colonization found in leaves of *E. oleracea* may be in part due to the size of the palm leaf in relation to the number of discs sampled; a palm leaf possesses a large area, thus reducing the probability of isolating dispersed fungal infection units present within the leaf.

Significant differences can be seen regarding the overall colonization frequencies of the endophytic fungi as influenced by leaf age, plant developmental stage and subsite. Although endophytic fungi were isolated from the three leaf age classes, the mature and the newly expanded leaves showed the greatest frequency of colonization during all four sampling periods (Fig. 1) indicating that older foliage hosts a large number of endophytes. The same trend was observed in leaves sampled from the Australian fan palm *L. ramsayi* (37). The increase in the colonization frequencies observed in the mature leaves could be explained by the following hypothesis: 1) the fungi were already present at an early stage in leaf development, and their mycelia gradually spread as the leaf expanded, older leaves would present higher inoculum densities, and therefore endophytes would be recovered more frequently; 2) arrival of new inoculum units. The high genetic diversity among the isolates revealed by isozyme analysis performed with isolates of *Xylaria* species (39) fully supports the assumption of a constant new infection. An increase of endophyte colonization with increasing age seems to be a common trend

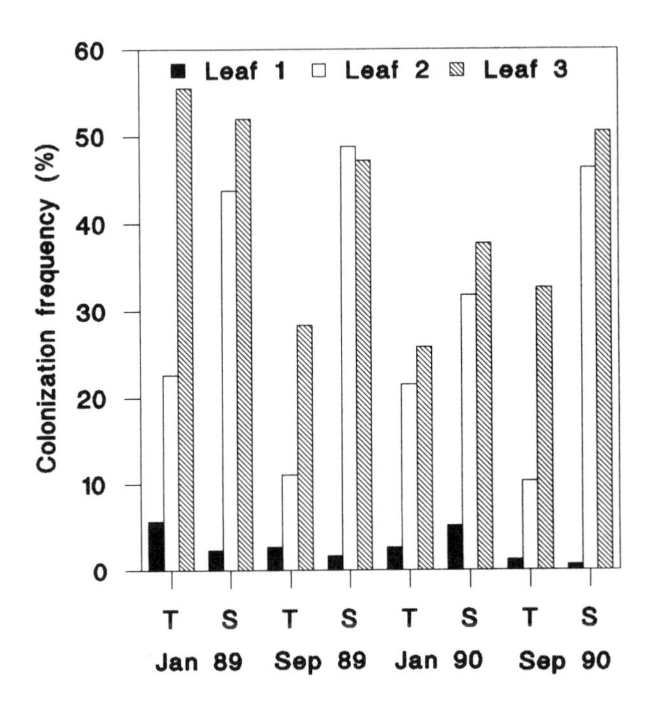

Fig. 1. Overall colonization frequency regarding age of leaves sampled from *Euterpe oleracea* during four sampling periods. The ANOVA indicated significant ($P < 0.001$) leaf-age effects. T= tree, S= sapling.

among endophytes, and has also been reported for species from temperate regions (2, 17, 20, 31).

In regard to the plant developmental stage, saplings showed higher colonization frequencies than trees (Fig. 2). The higher frequency of colonization found in saplings would be explained by the different proportion of sampled tissues, i. e. a sapling leaf represented a relatively great proportion of the leaf biomass, while a tree leaf represented only a small fraction of a single frond.

In general, there was a greater frequency of isolation of individual species found in plants collected in the high várzea than from those of the

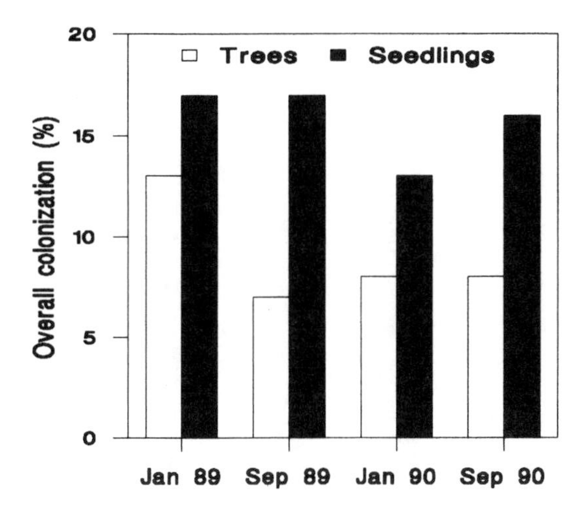

Fig. 2. Overall frequency of colonization (%) of endophytes in trees and saplings during four sampling periods. Differences detected in the colonization frequencies were significant ($P < 0.001$) according to the ANOVA.

low várzea (Fig. 3). This pattern was linked with the significantly high frequency of *Xylaria cubensis* (Mont.) Fr. and other xylariaceous fungi isolated from samples originated from the high várzea. High várzea is rarely subjected to flooding, in contrast to the low várzea, and it supports a large diversity of woody plants. Consequently, more substrata are available for *Xylaria* inoculum.

Distribution patterns

A low diversity of leaf endophytes was recorded in the fan palm, as only twelve species of endophytes were isolated. *Idriella licualae* K. F. Rodrigues & Samuels was the dominant endophyte and was present in 41% of the total number of discs infected. The second taxon most commonly isolated was *Xylaria cubensis*, accounting for 33%. The endophytic diversity recorded in *E. oleracea* was apparently higher. However, as the survey undertaken with leaves of the fan palm was based on a small sample, no direct comparison can be made. Fifty-seven species and five familial taxa were isolated from *E. oleracea* leaves (Table 1). Only two species, however, were dominant. *X. cubensis* was the most frequent and ubiquitous endophyte (21%) in leaves of *E. oleracea* followed by a loculoascomycete genus called cf. *Letendraea* sp.

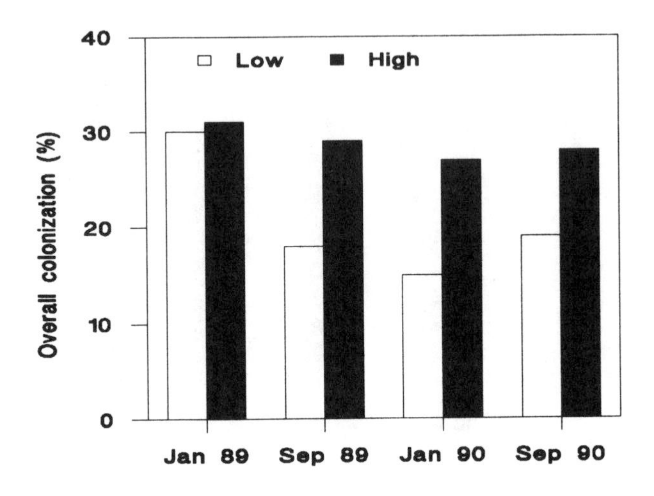

Fig. 3. Overall frequency of colonization (%) of endophytes in each collecting site (low and high várzea) during four sampling periods.

(15%). These cf. *Letendraea* isolates, however, differ from *Letendraea* Sacc. by the following characteristics: non stromatic, thin-walled, inostiolate ascomata; a subglobose to broadly clavate asci with a truncate base; submedially septate yellow brown ascospores; and a non fungicolous habit. A detailed taxonomic study of this taxon is currently being undertaken.

The endophytic communities in leaves of *E. oleracea* presented distinct patterns of species composition and frequency. Three groups of endophytes could be distinguished. The first was characterized by two species present in large amounts throughout all sampling times (*X. cubensis*, and cf. *Letendraea* sp.), the second by species consistently isolated at lower frequency (representing 48% of the total number of discs infected), and the third, the remaining species that were sporadically or rarely isolated, representing 16%.

Species that fall into the second group, in descending order of frequency, are *Xylaria arbuscula*, *Hypoxylon serpens*, *Thozetella* sp., *X. curta*, *X. adscendens*, *H. stygium*, *Phomopsis* sp. 1, Xylariaceae I, Xylariaceae III, Xylariaceae II, *Idriella euterpes*, *X.* cf. *multiplex*, *Colletotrichum gloeosporioides*, Xylariaceae IV, Xylariaceae V, arthroconidial fungus, *Phoma* sp., *Ustulina deusta*, and *Acrodictys elaeidicola*.

Table 1. Relative frequency of isolation (%) of most frequently occurring endophytic fungi in leaves from trees and saplings of *Euterpe oleracea*. Figures for endophytes refer to the number of discs colonized by a given fungus divided by the total number of discs infected, expressed as percentages; n= total number of discs infected, ana= anamorph.

Taxa	Trees (n=1,170)	Saplings (n=2,030)	Total (n=3,200)
Acrodictys elaeidicola M. B. Ellis	0.5	1.1	1.0
Arthroconidial	2.8	0.6	1.4
Colletotrichum gloeosporioides (Penz.) Penz. & Sacc.	2.2	1.6	1.8
Hypoxylon serpens (Pers.:Fr.) Kickx (ana)	4.2	4.8	4.6
H. stygium (Lév.) Sacc. (ana)	3.2	3.4	3.3
Idriella euterpes K. F. Rodrigues & Samuels	2.6	1.5	1.9
cf. *Letendraea* sp.	3.7	21.7	15.0
Phoma sp.	2.0	0.7	1.2
Phomopsis sp. 1	5.5	1.6	3.0
Thozetella sp.	0.6	5.8	3.9
Ustulina deusta (Hoffm.: Fr.) Petrak (ana)	0.7	1.4	1.1
Xylaria adscendens (Fr.) Fr. (ana)	5.7	2.4	3.6
X. arbuscula Sacc. (ana)	5.1	4.8	4.9
X. cubensis (Mont.) Fr. (ana)	24.6	19.0	21.0
X. curta Fr. (ana)	3.3	4.1	3.8
X. cf. *multiplex* (Kunze) Fr. (ana)	2.3	1.6	1.8
Xylariaceae I (ana)	1.6	3.8	2.9
Xylariaceae II (ana)	1.7	2.5	2.2
Xylariaceae III (ana)	2.5	3.2	2.9
Xylariaceae IV (ana)	1.8	1.7	1.7
Xylariaceae V (ana)	1.9	1.3	1.5

Note: Rare isolates included: *Acremonium* sp., *Anthostomella* sp., Basidiomycete, *Calonectria* sp., *Chloridium ? preussii* W. Gams & Hol.-Jech., *Colletotrichum* sp., *Curvularia pallescens* Boedijn, *Dendrodochium* sp., *Fusarium oxysporum* Schlecht., *F. sachari* (Butler) v. *elongatum* Nirenberg, *F. semitectum* Berk. & Rav.v. *majus* Wollenw., *F. verticillioides* (Sacc.) Nirenberg, *Graphium* sp., *Hypoxylon quisquiliarum* (Mont.) Mont.(ana), *Hypoxylon* sp. (ana: *Virgariella*-like), *Idriella amazonica* K. F. Rodrigues & Samuels, *I. asaicola* K. F. Rodrigues & Samuels, *Physalacria* sp., *Daldinia eschscholzii* (Ehrenb.) Rehm (ana), *Lasiodiplodia theobromae* (Pat.) Griff. & Maubl., *Leiosphaerella cocoes* (Petch) Samuels & Rossman, Mycelia Sterilia, *Mycoleptodiscus* sp., *Neosartoria* sp., *Nigrospora sphaerica* (Sacc.) E. Mason, *Nodulisporium* sp., *Oxydothis poliothea* Syd., *Penzigia ? berteri* (Mont.) Mill. (ana), *Pestalotiopsis palmarum* (Cooke) Steyaert, *Phomatospora* sp., *Phomopsis* sp. 2, *Physalacria* sp., *Physalospora* sp., *Trichoderma* sp., *Wardomyces* sp., *Xylaria allantoidea* (Berk.) Fr. (ana), *X.* cf. *anisopleura* (Mont.) Fr. (ana), *X.* cf. *castorea* Berk. (ana), *X. coccophora* Mont.(ana),*X.* cf. *microceras* (Mont.) Fr. (ana), *X.* cf. *obovata* (Berk.) Fr. (ana), *X.* cf. *palmicola* Winter (ana), X. cf. *telfairii* Berk. & Fr. (ana).

Endophytic taxa

The majority of the organisms isolated from *E. oleracea* ranges from potential saprobes to latent pathogens. In a search for the occurrence of fungal reproductive structures on falling fronds in the field as well as on herbarium specimens of *E. oleracea* deposited in the herbarium of the New York Botanical Garden, the same species of *Idriella*, *Physalospora*, and *Thozetella* were found that colonized internal tissues of apparently healthy leaves. Such observations suggest that these saprobes are involved with the process of leaf senescence, i. e., they are dormant in attached leaves and they possibly only start to fruit after the death of the leaf. *Colletotrichum gloeosporioides*, *Fusarium oxysporum*, *F. semitectum* var. *majus*, *Lasiodiplodia theobromae*, *Pestalotiopsis palmarum*, and *Ustulina deusta* have also been isolated from symptomless leaves of *E. oleracea*. These fungi are known to cause diseases in palms and other tropical plants (11, 21, 27, 34, 44, 47). The occurrence of fungal pathogens in apparently healthy plant tissues has been previously reported from a variety of plants (45) and has been discussed by Petrini (30).

Xylariaceous anamorphs

As a group, the Xylariaceae was more highly encountered in living leaves of *E. oleracea* than any other family. *Xylaria cubensis* was the dominant species, recovered in large numbers and consistently isolated over the four samplings. *X. cubensis* is a common, cosmopolitan, lignicolous species (40). Although *X. cubensis* is infrequently encountered as an endophyte, its endophytic host and geographic distribution is wide [a gymnosperm *Chamaecyparis thyoides* (L.) B. S. P. (4) and in the palms *L. ramsayi* (37) and *E. oleracea* (36)], reflecting the broad range of its non-endophytic, stromatic occurrence (40). Because of this, it is likely that this species has a much wider endophytic presence and significance than is now recognized. Other species of *Xylaria* were also frequently isolated. Among them were species belonging to the *X. polymorpha* species complex (41) such as *X. anisopleura*, *X. curta*, and others considered to be related to the *X. multiplex* group (7) - *X. adscendens*, *X. arbuscula*, *X. coccophora*, *X. microceras*, and *X. multiplex*.

Due to the regular recovery of xylariaceous fungi from living plant materials, they have been considered organisms well-adapted to an endophytic life (28, 35, 37). Despite the high frequency in which they are isolated as endophytes, their identification has proved to be difficult because teleomorphs are rarely formed in pure cultures. Endophytic Xylariaceae cultures must be compared with cultures derived from ascospores of known teleomorphs for a correct identification (28) and even then morphology alone is often not reliable.

Conclusions

The role of these endophytes within palm communities is still unknown. Benefits to the host plant such as antagonism towards pathogenic fungi or decreased susceptibility to phytophagous insects could be speculated. However, only by conducting studies with the dominant endophytic species, can these questions be answered. Investigations on the interactions of *E. oleracea* and its endophytes would be the next direction for future research, with *Xylaria cubensis* and cf. *Letendraea* as the most appropriate candidates for such ecological and physiological studies.

Based on endophytic studies in the temperate regions, where some fungal endophytes have shown a great potential use in biological control (30), it is expected that surveys undertaken in tropical regions also will turn in the same direction. The use of endophytic fungi to aid agriculture by means of biocontrol agents against pests, diseases and weeds has been discussed by Petrini (30). The knowledge of the endophytic fungi involved in naturally occurring resistance to insect predation or plant disease becomes a possibility for exploitation by turning them to economic use in palms or other tropical crops.

Acknowledgements

This research was supported by National Science Foundation Dissertation Improvement Grant BSR 8914564 to the New York Botanical Garden.

I thank Drs. Orlando Petrini and Gary J. Samuels for their assistance and valuable comments on the manuscript.

Literature Cited

1. Ashton, P. S. 1990. Species richness in tropical forests. Pages 239-251 in: Tropical Forests: Botanical Dynamics, Speciation and Diversity. L. B. Holm-Nielsen, I. C. Nielsen, and H. Balslev, eds. Academic Press. 380 pp.
2. Bertoni, M. D., and Cabral, D. 1988. Phyllosphere of *Eucalyptus viminalis*. II: distribution of endophytes. Nova Hedwig. 46: 491-502.
3. Bertoni, M. D., and Cabral, D. 1991. *Ceratopycnidium accharidicola* sp. nov., from *Baccharis coridifolia* in Argentina. Mycol. Res. 95: 1014-1016.
4. Bills, G. F., and Polishook, J. D. 1992. Recovery of endophytic fungi from *Chamaecyparis thyoides*. Sydowia 44: 1-12.
5. Boddy, L., and Griffith, G. S. 1989. Role of endophytes and latent invasion in the development of decay communities in sapwood of angiospermous trees. Sydowia 41: 41-73.

6. Cabral, D., Bertoni, M. D., and Varsavsky, E. 1990. Presence of endophytes in *Baccharis coridifolia* plants. Abstracts, p. 270. Fourth International Mycological Congress, Regensburg, Germany.
7. Callan, B. E. 1988. Cultural and anamorphic features of tropical Xylarias and related Xylariaceae. Ph.D. thesis, Washington State University. 127 pp.
8. Carroll, G. C. 1986. The biology of endophytism in plants with particular reference to woody perennials. Pages 205-222 in: Microbiology of the Phyllosphere. N. J. Fokkema, and J. van den Heuvel, eds. Cambridge University Press, Cambridge.
9. Carroll, G. C. 1991. Fungal associates of woody plants as insect-antagonists in leaves and stems. Pages 253-271 in: Microbial Mediation of Plant-Herbivore Interactions. P. Barbosa, V. A. Krischik, and C. G. Jones, eds. John Wiley and Sons, New York.
10. Carroll, G. C. and Carroll, F. E. 1978. Studies on the incidence of coniferous needle endophytes in the Pacific Northwest. Can. J. Bot. 56: 3034-3043.
11. Chase, A. R. 1991. Anthracnose (*Colletotrichum* Leaf Spot). Pages 2-3 in: Diseases and Disorders of Ornamental Palms. A. R. Chase, and T. K. Broschat, eds. American Phytopathological Society, St. Paul, MN.
12. Clay, K. 1986. Grass endophytes. Pages 188-204 in: Microbiology of the Phyllosphere. N. J. Fokkema, and J. van den Heuvel, eds. Cambridge University Press, Cambridge.
13. Clay, K. 1988. Fungal endophytes of grasses: a defensive mutualism between plants and fungi. Ecol. 69: 10-16.
14. Clay, K., Hardy, T. N., and Hammond, A. M., Jr. 1985. Fungal endophytes of grasses and their effects on an insect herbivore. Oecologia 66: 1-6.
15. De Bary, A.. 1866. Morphologie und Physiologie der Pilze, Flechten und Myxomyceten. Engelmann, Leipzig. 316 pp.
16. Dreyfuss, M. M., and Petrini, O. 1984. Further investigations on the occurrence and distribution of endophytic fungi in tropical plants. Bot. Helv. 94: 33- 40.
17. Espinosa-Garcia, F. J., and Langenheim, J. H. 1990. The leaf fungal endophytic community of a coastal redwood population. Diversity and spatial patterns. New Phytol. 116: 89-98.
18. Faifer, G. C., and Bertoni, M. D. 1988. Interactions between epiphytes and endophytes from the phyllosphere of *Eucalyptus viminalis*. III. Nova Hedwig. 47: 219-229.
19. Fisher, P. J., and Petrini, O. 1992. Fungal saprobes and pathogens as endophytes of rice (*Oryza sativa* L.). New Phytol. 120: 137-143.
20. Fisher, P. J., Anson, A. E., and Petrini, O. 1986. Fungal endophytes in *Ulex europaeus* and *Ulex gallii*. Trans. Br. Mycol. Soc. 86: 153-193.
21. Holliday, P. 1980. Fungus diseases of tropical crops. Cambridge University Press. Cambridge. 607 pp.

22. Jarvis, B. B., Mokhtari-Rejali, N., Schenkel, E. P., Barros, C. S., and Matzenbacher, N. I. 1991. Tricothecene mycotoxins from Brazilian *Baccharis* species. Phytochemistry 30: 789-797.
23. Junk, W. J. 1984. Ecology of the várzea floodplains of Amazon white-water rivers. Pages 271-293 in: The Amazon: Limnology and Landscape Ecology of a Mighty Tropical River and its Basin. H. Sioli, ed. Dr. W. Junk Publishers, Dordrecht.
24. Latch, G. C. M., and Christensen, M. J. 1982. Ryegrass endophyte, incidence, and control. N. Z. J. Agric. Res. 25: 443-448.
25. Latch, G. C. M., Christensen, M. J., and Samuels, G. J. 1984. Five endophytes of *Lolium* and *Festuca* in New Zealand. Mycotaxon 20: 535-550.
26. Latch, G. C. M., Hunt, W. F., and Musgrave, D. R. 1985. Endophytic fungi after growth of perennial ryegrass. N.Z. J. Agric. Res. 28: 165-168.
27. Ohr, H. D. 1991. *Fusarium* wilt. Pages 11-12 in: Diseases and Disorders of Ornamental Palms. A. R. Chase and T. K. Broschat, eds. American Phytopathological Society, St. Paul, MN.
28. Petrini, L. E., and Petrini, O. 1985. Xylariaceous fungi as endophytes. Sydowia 38: 216-234.
29. Petrini, O. 1986. Taxonomy of endophytic fungi of aerial plant tissues. Pages 175-187 in: Microbiology of the Phyllosphere. N. J. Fokkema, and J. van den Heuvel, eds. Cambridge University Press, Cambridge.
30. Petrini, O. 1991. Fungal endophytes of tree leaves. Pages 179-197 in: Microbial Ecology of Leaves. J. H. Andrews, and S. S. Hirano, eds. Springer-Verlag, New York.
31. Petrini, O., and Carroll, G. C. 1981. Endophytic fungi in foliage of some Cupressaceae in Oregon. Can. J. Bot. 59: 629-636.
32. Petrini, O., and Dreyfuss, M. 1981. Endophytische pilze in epiphytischen Araceae, Bromeliaceae und Orchidaceae. Sydowia 34: 135-148.
33. Philipson, M. N. 1989. A symptomless endophyte of ryegrass (*Lolium perenne*) that spores on its host - a light microscope study. N. Z J. Bot. 27: 513-519.
34. Ram, C. 1990. Comportamento de híbridos do coqueiro à *Botryodiplodia theobromae*, no estádio vegetativo em campo. Fitopatol. Bras. 15: 248-249.
35. Rodrigues, K. F. 1991. Fungos endofíticos em *Euterpe oleracea* Mart., com ênfase em Xylariaceae. Bol. Mus. Para. Emílio Goeldi, sér. Bot. 7 (2): 429-439.
36. Rodrigues, K. F. 1992. Endophytic Fungi in the Tropical Palm *Euterpe oleracea* Mart. Ph. D. thesis, City University of New York. 258 pp.
37. Rodrigues, K. F., and Samuels, G. J. 1990. Preliminary study of endophytic fungi in a tropical palm. Mycol. Res. 94: 827-830.
38. Rodrigues, K. F., and Samuels, G. J. 1992. *Idriella* species endophytic in palms. Mycotaxon 43: 271-276.

39. Rodrigues, K. F., Leuchtmann, A., and Petrini, O. 1993. Endophytic species of *Xylaria*: cultural and isozymic studies. Sydowia 45: 116-138.
40. Rogers, J. D. 1984. *Xylaria cubensis* and its anamorph *Xylocoremium flabelliforme*, *Xylaria allantoidea*, and *Xylaria poitei* in continental United States. Mycologia 76: 912-923.
41. Rogers, J. D. 1985. Anamorphs of *Xylaria*: taxonomic considerations. Sydowia 38: 255-262.
42. Sherwood-Pike, M., Stone, J. K., and Carroll, G. C. 1986. *Rhabdocline parkeri*, a ubiquitous foliar endophyte of Douglas-fir. Can. J. Bot. 64: 1849-1855.
43. Siegel, M. R., and Schardll, C. L. 1991. Fungal endophytes of grasses: Detrimental and beneficial associations. Pages 198-221 in: Microbial Ecology of Leaves. J. H. Andrews, and S. S. Hirano, eds. Springer-Verlag, New York.
44. Simone, G. W. 1991. *Pestalotiopsis* leaf spot. Page 20 in: Diseases and Disorders of Ornamental Palms. A. R. Chase, and T. K. Broschat, eds. American Phytopathological Society, St. Paul, MN.
45. Sinclair, J. B. 1991. Latent infection of soybean plants and seeds by fungi. Plant Dis. 75: 220-224.
46. Sioli, H. 1975. Tropical river: the Amazon. Pages 461- 487 in: River Ecology. B. A. Whitton, ed. University of California Press, Berkeley.
47. Turner, P. D. 1981. Oil Palm Diseases and Disorders. Oxford University Press, Oxford. 280 pp.
48. Varsavsky, E., Cabral, D., and Bertoni, M. D. 1990. Presence of macrocyclic trichothecenes in plants and endophytes of *Baccharis coridifolia*. Page 309, Abstracts, Fourth International Mycological Congress, Regensburg, Germany.
49. White, J. F., Jr., Morrow, A. C., and Morgan-Jones, G. 1990. Endophyte-host associations in forage grasses. XII. A fungal endophyte of *Trichachne insularis* belonging to *Pseudocercosporella*. Mycologia 82: 218- 226.

CHAPTER 7

MORPHOLOGICAL AND PHYSIOLOGICAL ADAPTATIONS OF BALANSIEAE AND TRENDS IN THE EVOLUTION OF GRASS ENDOPHYTES

J. F. White, Jr. and G. Morgan-Jones

Department of Biology
Auburn University,
Montgomery, AL 36117

Department of Plant Pathology
Auburn University
Auburn, AL 36849

Abstract

Coevolution of Balansieae and Poaceae is proposed. The earliest evolved grass-infecting Balansieae are suggested to parasitize bambusoid hosts, while more recently evolved Balansieae associate with pooid hosts. The epiphytic condition is suggested to be a primitive feature, while endophytism is considered advanced. Trends in colonization of endophytic niches become evident when comparisons of *Atkinsonella hypoxylon*, *Balansia epichloë*, and *Epichloë*/*Acremonium* spp. are made. *Atkinsonella hypoxylon* is epiphytic and possesses a more complex morphology and physiology than endophytic *Epichloë*/*Acremonium* species, suggesting a trend toward simplification as organisms colonized the relatively protected interior of grasses. *Balansia epichloë* is largely endophytic, except in early stroma development, which involves a brief epiphytic period of growth during which the fungus derives energy for development from leaf surface cuticular waxes. *Epichloë* and related *Acremonium* anamorphs are incapable of utilizing host waxes as energy sources, instead deriving energy from sugars or other energy-containing molecules that leak from host cells. A mechanism by which Balansieae regulate development on hosts depending on concentration of certain sugars is described.

The Balansieae is one of three tribes within the ascomycete subfamily Clavicipitoideae [Clavicipitaceae]. The other tribes are Clavicipiteae, containing *Claviceps* spp., and Ustilaginoideae, containing *Ustilaginoidea* spp. (9). The Balansieae comprize teleomorph genera *Atkinsonella* Diehl, *Balansia* Speg., *Balansiopsis* Höhnel, *Epichloë* (Fr.) Tul., and *Myriogenospora* Atk.; and anamorph genera *Acremonium* Link (section *Albo-lanosa* Morgan-Jones and W. Gams) and *Ephelis* Sacc. (4,5,9,16,17,19,28,30). Interest in the Balansieae is increasing due to a realization of the ecological and economic significance of some endophytic members of the group, particularly *Acremonium coenophialum* Morgan-Jones & W. Gams and *A. lolii* Latch, Christensen & Samuels. These endophytes have been shown to give grasses increased insect resistance (5), drought tolerance (27), and to cause toxic syndromes in cattle that consume infected grasses (1). Because of the growing interest in this group of fungi, there is an urgency in gaining knowledge of systematic and biological aspects of Balansieae and their relationships with host species. Recent molecular research on some Balansieae confirms close relatedness of its members and heralds impending opportunities to evaluate evolutionary relationships among species classified in the group (21). This task may be accomplished most effectively by reliance on a combination of molecular, morphological, physiological, and ecological criteria. Reliance on any single feature to the exclusion of other aspects of the organisms would likely result in failure to gain a complete appreciation of evolutionary relationships. In this paper an attempt is made to synthesize current information on adaptations and trends in morphology and physiology of several Balansieae so that a more complete picture of possible evolutionary relationships in the group might eventually be assembled.

Coevolutionary Relationships of Balansieae and Poaceae

Evolutionary relationships among the hosts of Balansieae are difficult to ascertain, but may give some indication of relationships among Balansieae (13,24,26). By correlating advancement of rust species (Uredinales) with their hosts, Savile (20) elucidated a chronology of evolution for several groups of Poaceae. It was noted that the most primitive rusts infect bambusoid grasses, while the most advanced are present on pooid grasses, with those of panicoid and chloridoid grasses being slightly less advanced than those of pooid hosts. Arundinoid grasses are infected by rusts whose features are intermediate in advancement between bambusoid and panicoid grasses. As a result of these correlations, it was postulated that the chronology of origin of grasses was bambusoid, arundinoid and andropogonoid, panicoid, chloridoid, and pooid groups. If the Balansieae have coevolved with grasses in a similar fashion to Uredinales, it should be feasible to determine a chronology of origin of Balansieae relative to the chronology of host grasses proposed by Savile (2,13,20). In this respect,

Balansia linearis (Rehm.) Diehl, *Balansiopsis gaduae* (Rehm.) Hohnel, and *B. pilulaeformis* (Berk. & Curt.) Diehl, all known to occur on bambusoids (9), may be among the oldest of the grass-infecting Balansieae. Whether they predate Balansieae, such as *Balansia cyperacearum* (Berk. & Curt.) Diehl and *B. cyperi* Edg., that parasitize sedges or other non-Poaceae is uncertain. Balansieae with arundinoid hosts, such as *Atkinsonella hypoxylon* (Pk.) Diehl and *A. texensis* (Diehl) Leuchtmann & Clay, that occur on *Danthonia* spp. and *Stipa leucotricha* Trin., respectively (9), may have arisen after bambusoid-infecting species. Following arundinoid-infecting Balansieae are those infecting panicoids and chloridoids, e.g., *Balansia epichloë* (Weese) Diehl, *B. strangulans* (Mont.) Diehl, and *B. subnodosa* (Atk.) Chardon (9). The *Epichloë/Acremonium* endophytes, infecting predominantly pooid grasses (8,28), are expected to be the most recently evolved.

General Treads in Host Relationships

Within the Balansieae, specific adaptive variations in fungus-host relationships, life cycles, and physiological capacities of mycobionts are evident and seem to correlate with the expected chronology of origin. An apparent trend is suggested in the progressive colonization of the internal environments of host grasses and establishment of endosymbiotic states. Concurrent with this trend is reduction in destructive effects on hosts, terminating, in the case of an endophyte such as *A. coenophialum*, in a totally benign condition. Alterations in fungus-host relationships involve changes in structural features, life cycles, and physiological capacities of individual species of Balansieae. The authors have not examined fresh material of bambusoid-infecting Balansieae. However, Diehl (9) reported severe developmental malformations of hosts and anomalous morphologies for these fungi. They may represent evolutionary 'prototypes' of the Balansieae. In the following discussion of trends and adaptations, emphasis will be placed on Balansieae that we have examined extensively, including some of the arundinoid-infecting Balansieae and those thought to be of more recent origin. Comparisons of *Atkinsonella hypoxylon*, *Balansia epichloë*, *Epichloë typhina* (Fr.) Tul., and *Acremonium coenophialum* are made (Table 1).

Atkinsonella hypoxylon exhibits what we consider to be a primitive condition among these species. It tends to remain epiphytic, a condition which might reasonably be expected as a precurser to endophytic development. Colonization of grasses by *A. hypoxylon* often results in destruction of 20-100% of the inflorescences on host individuals due to formation of stromata on developing inflorescences (30; Fig. 1). Concurrent with reduced fecundity, infected plants demonstrate enhanced vegetative growth, perhaps due to the diversion of energy from inflorescence formation

TABLE 1. Selected morphological and physiological features of some Balansieae[a]

Features	*A. hypoxylon*	*B. epichloë*	*E. typhina*	*A. coenophialum*
Frequency of stromata formation	high	high	low to moderate	never
Location of stromata	culms	leaves	culms	--------
Formation of microconidia on host	+[b]	-	+	-
Formation of macroconidia on host	+	+	-	-
Formation of ascospores on hosts	+	+	+	-
Hydrolysis of waxes	+	+	-	+
Hydrolysis of proteins	+	+	+	+
Hydrolysis of starches	-	+	+	+
Hydrolysis of oils	+	+	+	?
Capacity to utilize simple sugars	high	low	high	low to moderate

[a]Data represent studies on two isolates each of *A. hypoxylon* from *D. spicata* and *B. epichloë* from *S. poiretii*, and numerous isolates of *E. typhina* from *Elymus canadensis* L. and *Agrostis hiemalis* (Walt.) B.S.P., and of *A. coenophialum* from *F. arundinacea*.

[b]+ indicates presence of feature and - indicates absence of feature.

to vegetative processes or alternatively growth regulator effects induced by the fungus (4,18). There is conflicting evidence concerning seed transmission of *A. hypoxylon*. One study suggests that seed transmission may occur when cleistogamous flowers are formed on grasses (7). However, investigations which we have conducted on populations of *Danthonia spicata* (L.) R. & S. suggests that seed transmission may not be a

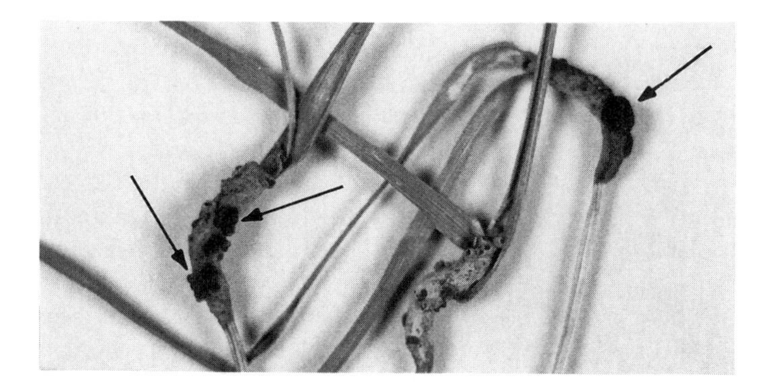

Fig. 1. Stromata of *Atkinsonella hypoxylon* on culms of *Danthonia spicata* [arrows indicate developing ascostromata].

Fig. 2. Black ascostroma of *Balansia epichloë* on upper surface of leaf of *Sporobolus poiretii*. Arrow indicates rolled region of an adjacent leaf containing a developing stroma.

regular occurrence and is probably not an important means of propagation in this fungus (White and Morgan-Jones, unpublished). Indeed, since *A. hypoxylon* is epiphytic throughout its life cycle, it is unlikely that seed transmission occurs regularly, as this process has been shown to involve endophytic development of mycelium within embryos (11,15,33).

Balansia epichloë and *Epichloë/Acremonium* species demonstrate advancements beyond the condition of *A. hypoxylon*, in that they are partially to entirely endophytic of host plants. In *B. epichloë*, mycelium is present in tissues of leaves but absent in stems and seeds. To propagate itself, the fungus must egress from developing leaves to form an ectophytic stroma on the upper surface of leaves (Fig. 2). It does not appear to do physical damage to host leaves and does not destroy seed producing capacity of plants since stromata occur on leaves rather than inflorescences. However, both growth reductions and enhancements have been reported to occur in infected hosts (6,9). Additionally, infected plants may form fewer fertile tillers than uninfected host individuals, perhaps due to nutrient drain on plants or effects of growth regulator compounds produced by the mycobionts (18).

Epichloë and related *Acremonium* endophytes show a highly decreased tendency to form stromata on plants, relying instead on seed transmission (32). Some of these endophytes [e.g., *A. coenophialum*] appear to exist entirely endophytic of hosts, without egressing from plants to form stromata or propagules of any type. The propensity of these endophytes for seed transmission may, in part, account for their widespread distribution in pooid grasses (8,28). In *Epichloë* it is often evident that damage to host reproductive capacity is minimized in that few infected plants actually bear stromata and frequently on plants with stromata only a small percentage of culms may bear them (29,32). It is apparent that plants bearing fungi that produce numerous stromata on inflorescences have decreased fecundity (3). In contrast, endophyte-infected plants on which stromata are rare or absent show both vegetative growth and seed production increases when compared to uninfected host individuals (6). Because of the widespread nature of the *Epichloë/Acremonium* group of endophytes and known beneficial effects in terms of increased hardiness, relationships between grasses and endophytes in this group are often considered mutualisms (5).

Morphological Adaptations of Balansieae

It is reasonable to suggest that as Balansieae adapted to the relatively stable internal environments of plants, aspects of life cycles formerly needed to adapt organisms to the dynamic external environments of plants were dispensed with and features adaptive to the internal environments were acquired (5,14). Consequently, variations in ecology and morphology that are evident among the Balansieae (Table 1) may reflect a trend toward endosymbiosis.

Atkinsonella hypoxylon produces three distinct spore states, a microconidial [*Acremonium*] state, a macroconidial [*Ephelis*] state, and a meiotically formed ascospore state. Propagule states appear to have distinct roles in the life cycle of the organism. Microconidia have been shown to function as spermatia and initiate formation of the ascospore state once transferred to a stroma of the opposite mating type (15,17). The development of ascostromata on masses of macroconidia (9,16) suggests that a possible function of these conidia may be to become structural components of the pulvinate ascostromata. This is supported by histological studies of developing ascostromata of *Atkinsonella texensis* which revealed that they are initiated from concentric layers of conidiogenous cells and macroconidia (17). Additionally, when stromata are fertilized with compatible microconidia, macroconidia develop at the site of inoculation (White and Morgan-Jones, unpublished). A macroconidial mass forms an expanding cushion that develops into the black ascostroma. It is possible that the apparently water-dispersed *Ephelis* conidia, in this instance, play a dual role and that they may also function as true propagules bringing about host infections, probably principally of tillers in the immediate vicinity of stromata. Ascospores are ejected into the air and probably serve as the primary propagules to initiate infections of distant susceptible host individuals (9).

Balansia epichloë lacks a microconidial state, but exhibits a macroconidial state which may develop entirely into a flattened stroma (Figs. 2,3). Whether ephelidial macroconidia may function as spermatia, in addition to serving as structural components of developing stromata, is still undetermined. In contrast, *Epichloë* possesses microconidia but lacks macroconidia (31). The microconidia function as spermatia, as was evident in *Atkinsonella*. However, macroconidia appear unnecessary in *Epichloë*, since ascostromata develop from a mycelium that spreads laterally on the stroma once a microconidium of the opposite mating type has been deposited (31). Some of the *Acremonium* endophytes show further reductions in life cycle in that they exist *in vivo* only in the form of an endophytic mycelium without production of propagules or sexual recombination (33). However, most of these will form microconidia *in vitro* on agar media.

Physiological Adaptations of Balansieae

The relationship between the Balansieae, discussed herein, and their host plants, may be described as continuous and developmental. They maintain a continuous symbiotic association with host plants as hosts develop through successive growth stages. This is in contrast to the association evident in the Clavicipiteae, where *Claviceps* spp. spend long periods of time independent of host plants as resistant sclerotia in soil, later infecting only florets of

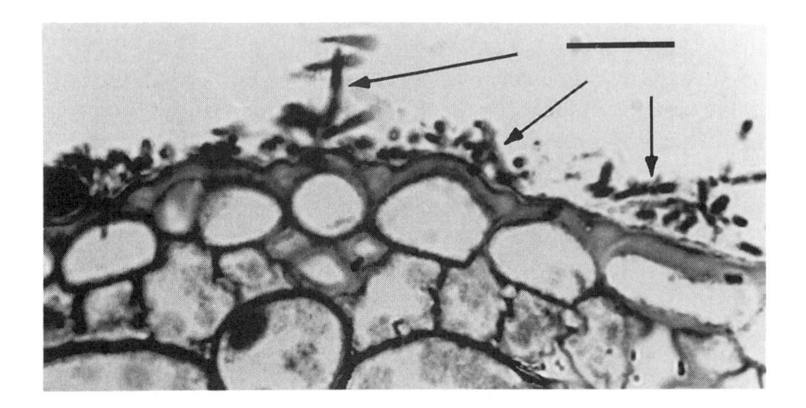

Fig. 3. Section of developing leaf, showing macroconidia of *Balansia epichloë* (arrows) adhering to cuticle of leaf [scale bar = 15 µm].

susceptible hosts (9). Some Balansieae may show a life cycle similar to *Claviceps* (e.g., *Balansia claviceps* Speg.), where stromata on inflorescences, termed "pseudomorphs" by Diehl (9), become resistant and may remain on dead culms or fall to the soil as apparent resting structures. Later, pseudomorphs germinate to form stipitate ascostromata nearly identical to those of *Claviceps* (9).

Continuous maintenance of a symbiotic relationship requires that Balansieae have available on host plants energy sources for sustenance during all developmental stages of the host. One such energy source is the waxes covering surfaces of leaves. Recent substrate utilization studies have demonstrated that *A. hypoxylon* and *B. epichloë* both possess capacity to utilize waxes, while *E. typhina* and *Acremonium* species lack that ability (30; Table 1). The continuous epiphytic growth of *A. hypoxylon* on plant surfaces (Fig. 4) may be supported, in part, from energy derived from the waxy cuticle covering grass plants. Other energy-containing compounds may be released from host tissues through the epidermis to the superficial mycelium. In *B. epichloë* epiphytic development occurs during early stroma development, when leaves become covered by macroconidia which adhere closely to the waxy cuticle (Fig. 3). This proliferation of macroconidia on leaf surfaces may be fueled by energy derived from waxes accumulating on leaves. Electron microscope studies of the fungus-host interface of *Sporobolus poiretii* (R. & S.) Hitchc., infected by *B. epichloë*, indicate that the cuticle is being hydrolyzed during stroma development

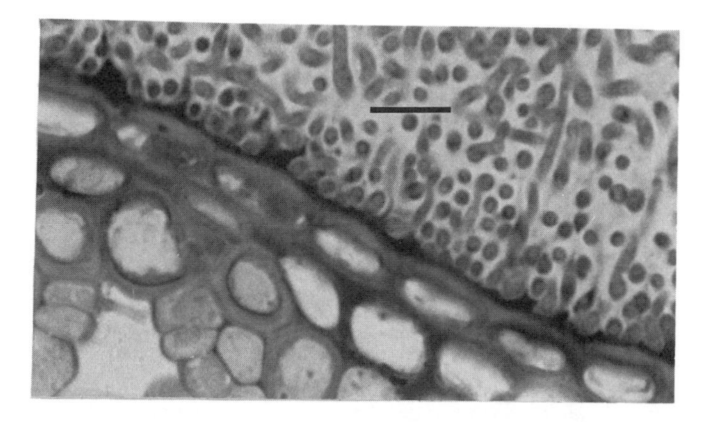

Fig. 4. Cross-section of stroma of *Atkinsonella hypoxylon* showing epiphytic development of mycelium concentrated on cuticle of leaf [scale bar = 10 µm].

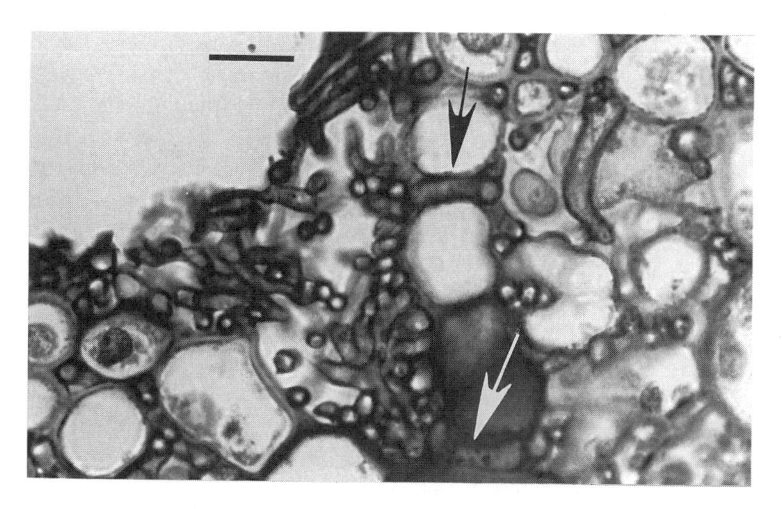

Fig. 5. Cross-section of young leaf of *Sporobolus poiretii* showing hyphal bridges (arrows) crossing epidermis in a longitudinal groove on upper surface of leaf [scale bar = 12 µm].

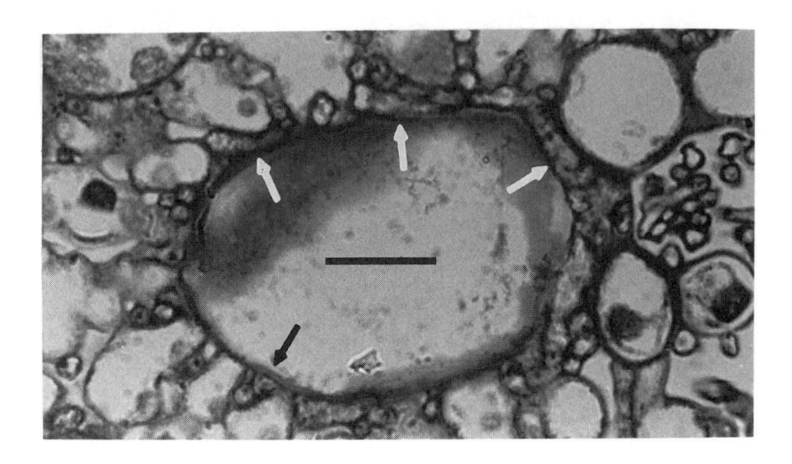

Fig. 6. Cross-section of leaf of *Sporobolus poiretii* showing concentrations of hyphae (arrows) surrounding a bulliform cell [scale bar = 7 μm].

(19,30). However, later stages of stroma development in *B. epichloë* are likely fueled by internal energy sources via 'hyphal bridges' (Fig. 5) which connect external stromatal mycelium with the internal network of intercellular hyphae. Hyphal bridges develop predominantly in proximity to bulliform cells which occur in sunken longitudinal grooves on upper surfaces of leaves of *S. poiretii* (Fig. 6). It is apparently the nutritional connection with this region that accounts for the formation of masses of macroconidia in rows above the bulliform cells on the conidial stroma (19). The bulliform cells function to roll and unroll leaves (26). Typically, where developing stromata are evident, leaves remain rolled in the region just surrounding a stroma (9; Fig. 2). This tendency to remain rolled is likely due to effects induced by *B. epichloë* on bulliform cells, retarding their ability to expand and expose stromata until they are fully developed.

Additional energy sources which may be used by Balansieae include lipids, proteins, and carbohydrates (Table 1). Isolates of *A. hypoxylon*, *B. epichloë*, *E. typhina*, and various *Acremonium* spp. were found to colonize lipid droplets *in vitro* (29), suggesting that they may gain some energy for development from lipids and oils present in plants. However, how they may gain access to these substances, which are contained within host cells, is unknown. Additionally, species of Balansieae have been shown to possess the capacity to degrade proteins when grown on litmus-milk agar or other milk protein-containing media (30,34; White and Morgan-Jones, unpublished data; Table 1). This capacity may provide, not only energy, but also nitrogen necessary for enzyme production by the fungi, or alternatively, it might be a means of deactivating host enzymes which are competing for energy-containing molecules for host anabolic activities like cell wall construction (10). Starch degradation *in vitro* has been demonstrated for *B. epichloë*, *E. typhina*, and *A. coenophialum* (Table 1) and suggests that these endophytes may draw on host storage carbohydrates for energy requirements. However, since endophytic hyphae of Balansieae are extracellular it is difficult to see how they may tap starch reserves which, like those of lipids, are housed internal of host cells (23). Whether fungal enzymes might act superficially on cell wall components or alternatively may be secreted into host cells to release energy containing molecules from storage starches is unknown. Regardless, the hydrolysis of starch by endophytic Balansieae, but not epiphytic *A. hypoxylon*, suggests that this capacity may be adaptive to the endophytic niche.

The main energy-yielding compounds for many Balansieae are probably simple sugars. One of the most commonly available sugars is sucrose, which is present in low levels in apoplastic spaces where some endophytes reside (10). Additionally, it seems probable that inflorescence-choking Balansieae, such as *A. hypoxylon* and *E. typhina*, may gain a substantial portion of their energy for early stromata development from this sugar which is expected to be abundant in the region of the meristematic inflorescence, to which it is translocated to supply energy and carbon for anabolic activities of the expanding inflorescence (35). To utilize sucrose, Balansieae secrete invertases, which cleave it to its component monomers glucose and fructose, that may then be absorbed individually by hyphae using membrane-bound permeases (10). Other sugars available to Balansieae include xylose and arabinose, both components of cell walls of grasses. The association of endophytes with meristematic cells (35), makes them well situated to utilize these sugars before they are polymerized into cell walls. Ultrastructural studies on several Balansieae (12,19,22) have not demonstrated cell wall degradation by proximally located hyphae, suggesting that once sugars are incorporated into cell walls they are unavailable to the endophytes.

Regulation of Fungal Development in Vivo

While sugars supply energy for growth on hosts, they have additional importance in providing a means by which the growth rates of some Balansieae are regulated. In this mechanism, mycelial growth rates are adjusted depending on types and concentrations of sugars present in the microenvironments of hyphae. This enables Balansieae to grow in specific locations of plants and adjust the rate at which mycelium proliferates within tissues. For example, *in vitro* studies have shown that isolates of *B. epichloë* may develop on low concentrations of glucose and some strains additionally show some growth on low concentrations [usually less than 1%

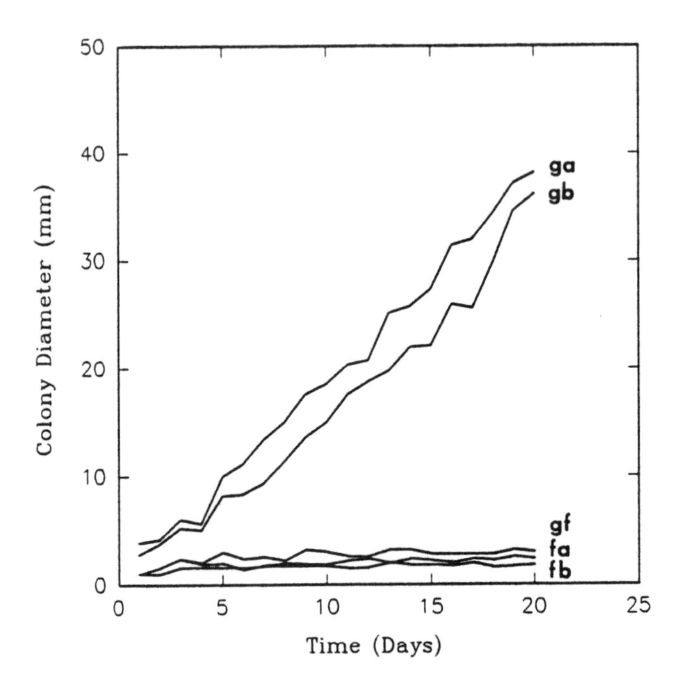

Fig. 7. Growth curves [25 C] for *Balansia epichloë* grown on media containing Murashige and Skoog's Salt Base (M.S.), 1.5% agar, and one of the following filter-sterilized sugars: 0.5% [wt./vol.] glucose (ga), 0.5% fructose (fa), 1% glucose (gb), 1% fructose (fb), and a mixture of 0.5% glucose and 0.5% fructose (gf).

wt./vol.] of fructose and arabinose, however higher concentrations of the latter two sugars are completely inhibitory to mycelial growth (30). While glucose is usually not inhibitory, when it is mixed with an inhibitory sugar, the mycelial growth rate is determined by the inhibitory sugar. Thus, growth curves (Fig. 7) of *B. epichloë* grown for 20 days at 25C on 0.5% and 1% concentrations of glucose, fructose, and the mixture of glucose/fructose [0.5% each] show that mycelial growth is completely inhibited on both concentrations of fructose and the glucose/fructose mixture, but only minimally inhibited on the higher concentration of glucose. The way that *B. epichloë* responds to high concentrations of certain sugars may explain why stromata develop on leaves rather than on inflorescences where high levels

Fig. 8. White conidial stroma of *Epichloë typhina* (arrow) on culm of *Agrostis hiemalis* (Walt.) B.S.P.

of many different sugars are expected to prevent mycelial development. The formation of stromata toward the middle of the leaf blade of *S. poiretii* may be the result of optimal levels of sugars in that region of leaves as they develop. After stromata have partially formed, the hyphal bridges to internal mycelium may provide low levels of sugars [e.g., apoplastic sucrose or glucose from starch degradation] for continued stromatal maturation (19,30).

In both *Atkinsonella* and *Epichloë*, where stromata develop on inflorescence primordia (Figs. 1,8), fungi show growth on a greater range of sugars and at higher concentrations (Figs. 9,10,11). Here, sensitivity of individual strains to sugar concentrations determines the rate that mycelium develops around and within expanding inflorescences and consequently, the percentage of stromata that are successfully formed (30). Strains that do not show growth reductions as sugar concentration is increased tend to form stromata on a majority of inflorescences produced on hosts, and may effectively sterilize them. In a recent study of *A. hypoxylon*, one strain [isolate 2], which formed stromata on all culms of its host was demonstrated to show only slight decreases in growth rate as sugar concentrations were increased (Fig. 9). Another strain [isolate 1], which produced stromata on only about 20% of the culms on infected plants, demonstrated considerable growth reduction as fructose concentration increased (Fig. 10).

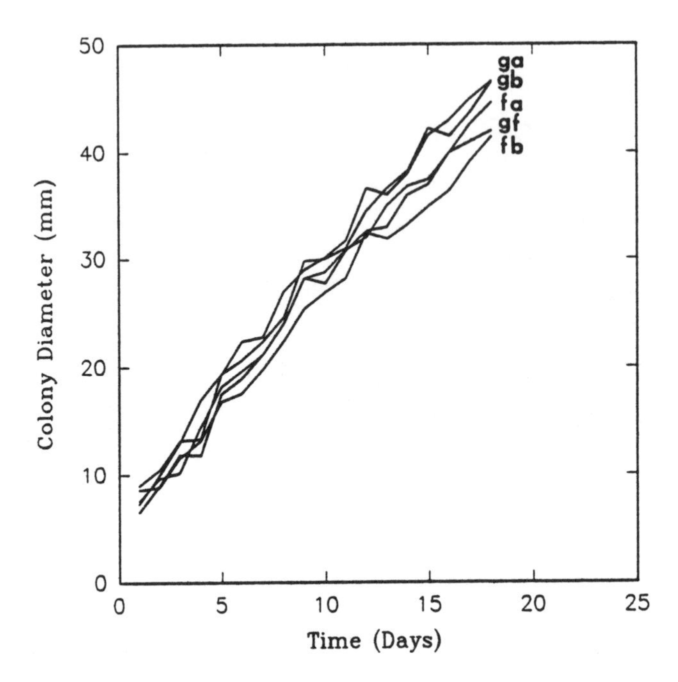

Fig. 9. Growth curves [25 C] for *Atkinsonella hypoxylon* [Isolate 2] on M.S. media containing 0.5% glucose (ga), 0.5% fructose (fa), 1% glucose (gb), 1% fructose (fb), and a mixture of 0.5% glucose and 0.5% fructose (gf).

Similar growth reduction occurs in *E. typhina* and related *Acremonium* endophytes. A typical growth curve is indicated in Fig. 11, where an *Acremonium starrii* White & Morgan-Jones from *Poa palustris* L. is shown to have a growth response to increasing sugar concentration that is similar to that observed for *A. hypoxylon* [isolate 1]. An important difference between *A. hypoxylon* and *E. typhina* is that mycelium of the latter is largely endophytic and during stromata formation proliferates within as well as superficial on host tissues of developing inflorescences and surrounding leaves. In the *Epichloë/Acremonium* group energy cannot be derived from waxes present on leaf surfaces, but exclusively from sugars or other compounds that are present external to meristematic cells or may be induced to 'leak' from those cells (30,34,35). As a result of this, mycelium of *E.*

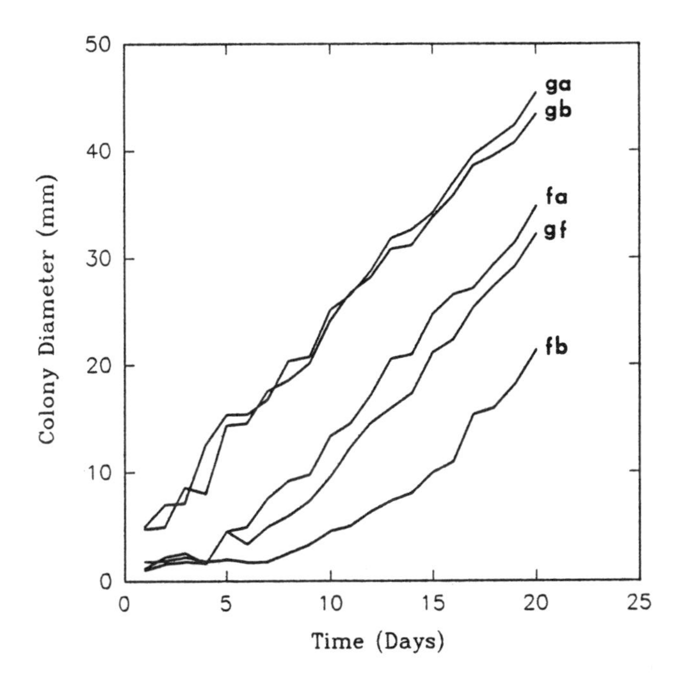

Fig. 10. Growth curves [25 C] for *Atkinsonella hypoxylon* [Isolate 1] on M.S. media containing 0.5% glucose (ga), 0.5% fructose (fa), 1% glucose (gb), 1% fructose (fb), and a mixture of 0.5% glucose and 0.5% fructose (gf).

typhina in early stromata formation emerges from young leaves and proliferates in spaces between leaves, where soluble sugars may be available (Fig. 12). Strains of the endophytes that show reductions in growth rate as concentrations of sugars fructose, xylose, and arabinose increase, also show marked decreases in the number of stromata that are successfully formed on plants (30,32). Because fructose inhibits development of endophytes, sucrose may also be regulatory. For each sucrose molecule that is lysed by fungal invertases, a fructose is released. As this sugar becomes abundant in the region of the mycelium in the meristematic inflorescence, it is expected to reduce the growth rate of that mycelium. Thus lower concentrations of sucrose may stimulate endophyte growth, but as concentration is increased, fructose inhibition may limit growth. This phenomenon is illustrated by

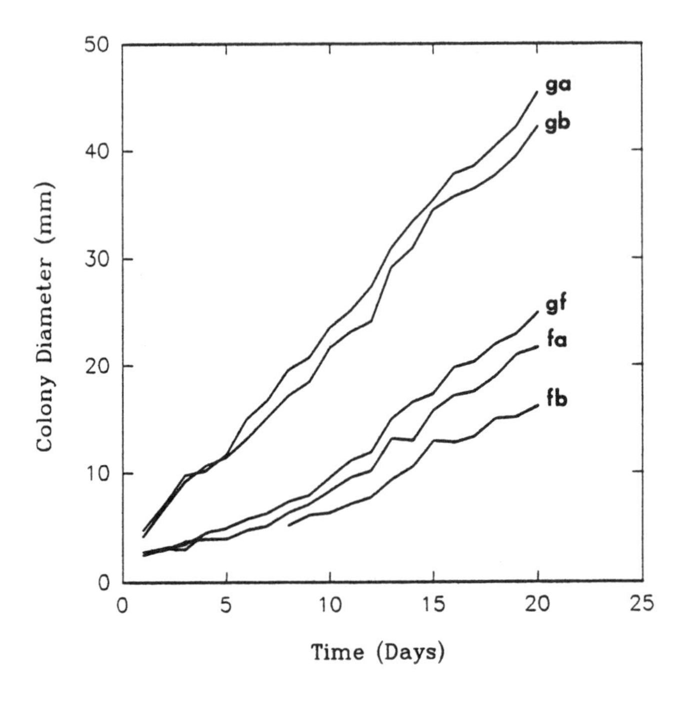

Fig. 11. Growth curves [25 C] for a nonstroma-forming *Acremonium* sp. [isolated from *Poa palustris*] grown on M.S. media containing 0.5% glucose (ga), 0.5% fructose (fa), 1% glucose (gb), 1% fructose (fb), and a mixture of 0.5% glucose and 0.5% fructose (gf).

growth curves for *Acremonium typhinum* Morgan-Jones & W. Gams [isolate Et-61] grown on increasing concentrations of filter-sterilized sucrose and glucose (Fig. 13). Here, increasing concentrations of glucose and sucrose are both seen to be stimulatory to growth at lower sugar concentrations, but as concentrations increase beyond 3.4%, marked inhibition becomes evident on sucrose, whereas only minimal growth reduction occurs on glucose. If an endophyte is suppressed from rapid proliferation in the inflorescence primordia, and thus prevented from completing stroma development, then it remains slowly growing in the inflorescence and is incorporated into differentiating carpels (34,35). Its next opportunity for sugar-stimulated proliferation likely occurs when the embryo differentiates, at which time the endophyte intercellularly invades the embryo in the seed and grows into the

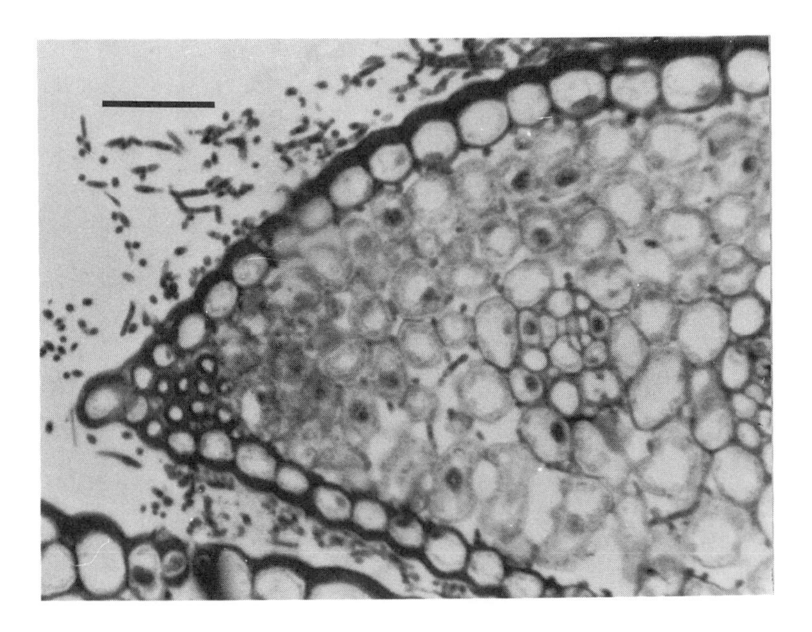

Fig. 12. Section of developing stroma of *Epichloë typhina* on *Lolium perenne* L., showing mycelial proliferation in spaces between leaves [scale bar = 15 μm].

meristematic coleoptile and embryonic leaves (33,34). The impetus for embryo infection is probably the release of sugars from the shoot meristem of the embryo (35).

The regulation of fungal development by sugar concentration seems to provide an explanation for certain ecological phenomena exhibited in the Balansieae. For example, it has recently been demonstrated that soil fertility level and stromata formation of *E. typhina* may be inversely related (25). When soil fertility is high the percentage of culms bearing stromata is low, and when soil fertility is low, stromata formation is high. This is expected in that high soil fertility results in an increased efficiency of all metabolic activities of plants and, of particular relevance, photosynthesis. Increased efficiency of photosynthesis results in greater production of sugars which are transported into the developing inflorescences to provide energy for flower and seed formation. Increased sugars in the inflorescences result in a decrease in growth rate of endophytic mycelium within inflorescences

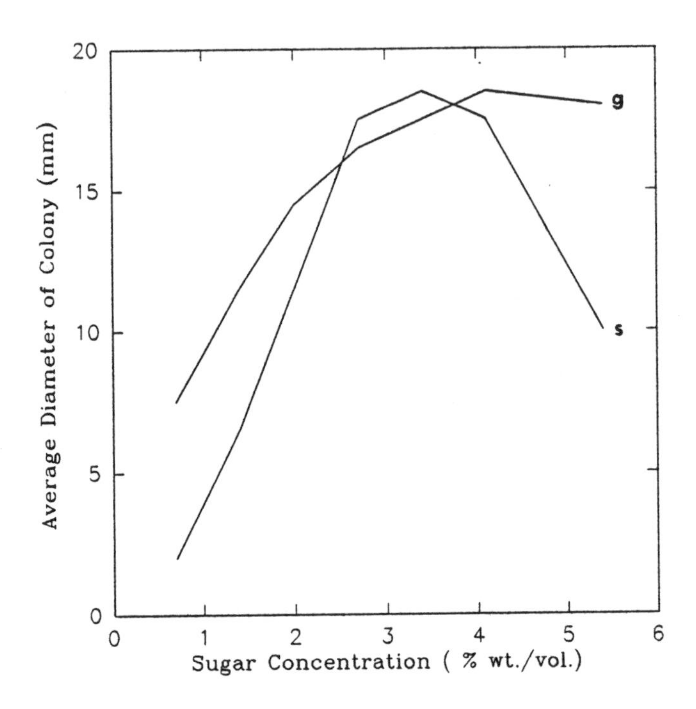

Fig. 13. Growth curves [25 C] for *Epichloë typhina* [Isolate Et-61] grown for 14 days on increasing concentrations of filter-sterilized glucose (g) and sucrose (s).

and failure of the endophyte to form stromata. On low fertility soils metabolic efficiency is reduced so that less sugar is translocated into inflorescences and sugar concentrations do not rise to inhibitory levels that prevent stromata development on culms.

During the process of stromata development in *E. typhina* mycelium proliferates intercellularly in the grass inflorescence primordia and young surrounding leaves (34), establishing close connections with numerous cells from which sugars may continue to be obtained for further maturation of stromata. Vascular tissues may be particularly important in supplying this continued energy for maturation. Sections of leaves and leaf sheaths imbedded in stromata of *E. typhina* and *B. epichloë* often reveal an abundance of intercellular hyphal strands among both xylem and phloem tissues (Fig. 14). Sugars released from this tissue may eventually supply the energy for development of perithecia and ascospores on stromata, a process

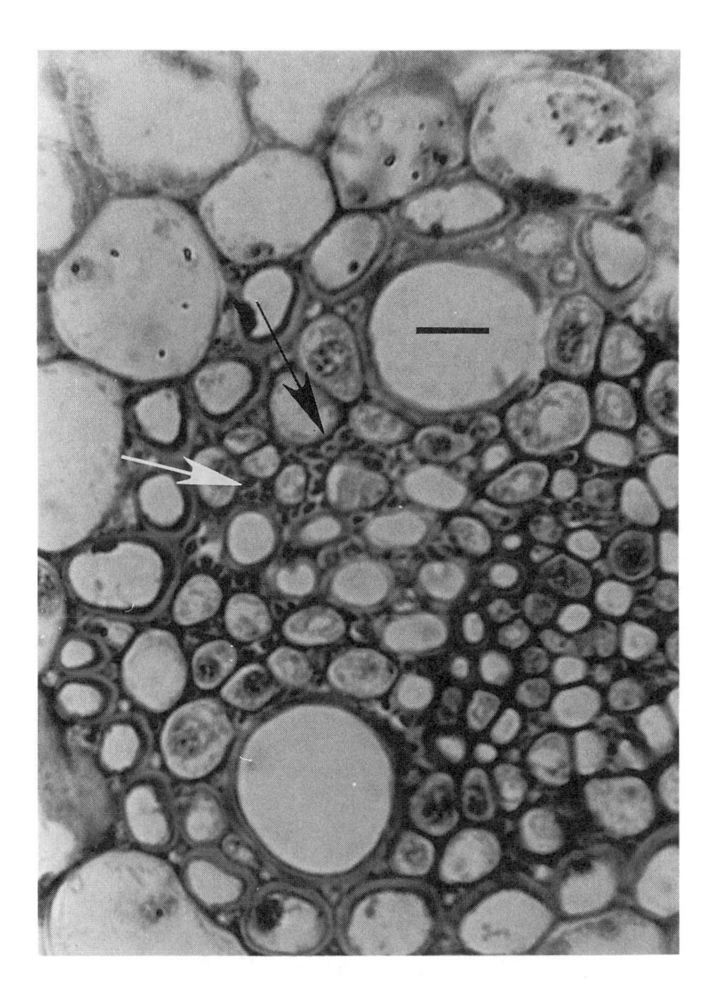

Fig. 14. Cross-section of leaf of *Lolium perenne* embedded in a stroma of *Epichloë typhina* showing intercellular hyphae (arrows) in vascular tissue [scale bar = 7 μm].

that requires weeks of development after stromata are formed (30). In this respect, it is significant that stromata develop on leaves, or leaf sheaths, since the photosynthetic activity of affected leaves likely supplies a substantial component of the fixed carbon used by the fungus for stroma maturation (10).

While it is apparent that Balansieae have coevolved with their hosts and certain trends are evident, there is much that is still unknown. The precise roles of a diverse array of fungus-produced enzymes, alkaloids, and perhaps other compounds requires elucidation. The mechanism by which Balansieae are regulated through sugar concentration is intriguing. It is of interest to determine whether this growth regulation method evolved in the Balansieae and is limited to certain members of the group or alternately occurs in other plant-fungus associations. It is probable that until we develop a more thorough understanding of all aspects of the biology of Balansieae and indeed the entire Clavicipitaceae, our attempts at elucidating evolutionary relationships of its members will be tentative.

Literature Cited

1. Bacon, C. W., Porter, J. K., Robbins, J. P., and Luttrell, E. S. 1977. *Epichloë typhina* from toxic tall fescue grasses. Appl. Environ. Microbiol. 34:576-581.
2. Barkworth, M. E., and Everett, J. 1986. Evolution in the Stipeae: identification and relationships of its monophyletic taxa. Pages 251-264 in: Grass Systematics and Evolution, T. R. Soderstrom, K. W. Hilu, C. S. Campbell, M. E. Barkworth, eds. International Grass Symposium, Smithsonian Inst., Washington, DC.
3. Bradshaw, A. D. 1959. Population differentiation in *Agrostis tenuis* Sibth. II. The incidence and significance of infection by *Epichloë typhina*. New Phytol. 58:310-315.
4. Clay, K. 1984. The effect of the fungus *Atkinsonella hypoxylon* (Clavicipitaceae) on the reproductive system and demography of the grass *Danthonia spicata*. New Phytol. 98: 165-175.
5. Clay, K. 1988. Clavicipitaceous fungal endophytes of grasses: coevolution and the change from parasitism to mutualism. Pages 79-105 in: Coevolution of Fungi with Plants and Animals, D. L. Hawksworth, and K. Pirozynski, eds. Academic Press, London.
6. Clay, K. 1990. Comparative demography of three graminoids infected by systemic, clavicipitaceous fungi. Ecology 71:558-570.
7. Clay, K. and Jones, J. P. 1984. Transmission of *Atkinsonella hypoxylon* (Clavicipitaceae) by cleistogamous seed of *Danthonia spicata* (Gramineae). Can. J. Bot. 62:2893-2895.
8. Clay, K. and Leuchtmann, A. 1989. Infection of woodland grasses by fungal endophytes. Mycologia 81:805-811.
9. Diehl, W. W. 1950. *Balansia* and the Balansieae in America. Agricultural Monograph 4. United States Department of Agriculture, Washington, DC. 82 pp.
10. Farrar, J. F., and Lewis, D. H. 1987. Nutrient relations in biotrophic infections. Pages 97-132 in: Fungal Infection of Plants, G. F. Pegg and P. G. Ayres, eds. Cambridge University Press, Cambridge.

11. Freeman, E. M. 1902. The seed-fungus of *Lolium temulentum*, L., the darnel. Philo. Trans. R. Soc. Lond. 196:1-29.
12. Hinton, D. M., and Bacon, C. W. 1985. The distribution and ultrastructure of the endophyte of toxic tall fescue. Can. J. Bot. 63:36-42.
13. Kellogg, E. A., and Campbell, C. S. 1987. Phylogenetic analysis of the Gramineae. Pages 310-322 in: Grass Systematics and Evolution, T. R. Soderstrom, K. W. Hilu, C. S. Campbell, and M. E. Barkworth, eds. International Grass Symposium, Smithsonian Inst., Washington, DC.
14. Law, R., and Lewis, D. H. 1983. Biotic environments and the maintenance of sex: some evidence from mutualistic symbioses. Biol. J. Linn. Soc. 20:249-276.
15. Leuchtmann, A., and Clay, K. 1989. Morphological, cultural and mating studies on *Atkinsonella*, including *A. texensis*. Mycologia 81:692-701.
16. Luttrell, E. S., and Bacon, C. W. 1977. Classification of *Myriogenospora* in the Clavicipitaceae. Can. J. Bot. 55: 2090-2097.
17. Morgan-Jones, G., and White, J. F., Jr. 1989. Concerning *Atkinsonella texensis*, a pathogen of the grass *Stipa leucotricha*: developmental morphology and mating system. Mycotaxon 35:455-467.
18. Porter, J. K., Bacon, C. W., Cutler, H. G., Arrendale, R. F., and Robbins, J. D. 1985. In vitro auxin production by *Balansia epichloë*. Phytochem. 24:1429-1431.
19. Rykard, D. M., Bacon, C. W., and Luttrell, E. S. 1985. Host relations of *Myriogenospora atramentosa* and *Balansia epichloë* (Clavicipitaceae). Phytopathology 75:950-956.
20. Savile, D. B. O. 1987. Use of rust fungi (Uredinales) in determining ages and relationships in Poaceae. Pages 168-178 in: Grass Systematics and Evolution, T. R. Soderstrom, K. W. Hilu, C. S. Campbell, and M. E. Barkworth, eds. International Grass Symposium, Smithsonian Inst., Washington, DC.
21. Schardl, C. L., Liu, J. S., White, J. F., Jr., Finkel, R. A., and Siegel, M. R. 1991. Molecular phylogenetic relationships of nonpathogenic grass mycosymbionts and clavicipitaceous plant pathogens. Pl. Syst. Evol. 178:27-41.
22. Siegel, M. R., Jarlfors, U., Latch, G. C. M., and Johnson, M. C. 1987. Ultrastructure of *Acremonium coenophialum*, *Acremonium lolii*, and *Epichloë typhina* endophytes in host and nonhost *Festuca* and *Lolium* species of grasses. Can. J. Bot. 65:2357-2367.
23. Siegel, M. R., Latch, G. C. M. and Johnson, M. C. 1987. Fungal endophytes of grasses. Annu. Rev. Phytopath. 25:293-315.
24. Smouter, H. and Simpson, R. J. 1989. Occurrence of fructans in the Gramineae (Poaceae). New Phytologist 111: 359-368.
25. Sun, S., Clarke, B. B., and Funk, C. R. 1991. Effect of fertilizer and fungicide applications on choke expression and endophyte transmission in chewings fescue. Pages 62-67 in: Proceedings Internat. Symposium

on *Acremonium*/Grass Interactions, S. S. Quisenberry and R. E. Joost, eds. Louisiana Agric. Exp. Sta., Baton Rouge.
26. Watson, L., Clifford, H. T., and Dallwitz, M. J. 1985. The classification of Poaceae: subfamilies and supertribes. Australian J. Bot. 33:433-484.
27. West, C. P., Oosterhuis, P. M., and Wullschleger, S. D. 1990. Osmotic adjustment in tissues of tall fescue in response to water deficit. Environ. Exp. Bot. 30:149-156.
28. White, J. F., Jr. 1987. The widespread distribution of endophytes in the Poaceae. Plant Disease 71:340-342.
29. White, J. F., Jr. 1988. Endophyte-host associations in forage grasses. XI. A proposal concerning origin and evolution. Mycologia 80:442-446.
30. White J. F., Jr., Breen, J. P., and Morgan-Jones, G. 1991. Substrate utilization in selected *Acremonium*, *Atkinsonella*, and *Balansia* species. Mycologia 83:601-610.
31. White, J.F., Jr., and Bultman, T. L. 1987. Endophyte-host associations in forage grasses. VIII. Heterothallism in *Epichloë typhina*. Amer. J. Bot. 74:1716-1721.
32. White, J. F., Jr., and Chambless, D. A. 1991. Endophyte-host associations in forage grasses. XV. Clustering of stromata-bearing individuals of *Agrostis hiemalis* infected by *Epichloë typhina* Amer. J. Bot. 78:527-533.
33. White, J.F., Jr., and Cole, G. T. 1986. Endophyte-host associations in forage grasses. IV. The endophyte of *Festuca versuta*. Mycologia 78:102-107.
34. White, J. F., Jr., Morgan-Jones, G., and Morrow, A. C. 1993. Taxonomy, life cycle, reproduction, and detection of *Acremonium* endophytes. Agriculture, Ecosystems and Environment 44:13-37
35. White, J. F., Jr., Morrow, A. C., Morgan-Jones, G., and Chambless, D. A. 1991. Endophyte-host associations in forage grasses. XIV. Primary stromata formation and seed transmission in *Epichloë typhina*: Developmental and regulatory aspects. Mycologia 83:72-81.

This research was supported by National Science Foundation grant (BSR-8922157).
The authors are grateful to Dr. R. J. Soreng [Cornell University] for discussions concerning evolutionary relationships in the Poaceae and Dr. C. L. Schardl [University of Kentucky, Lexington] for correspondence concerning possible evolutionary scenarios in the Balansieae.
Alabama Agricultural Experiment Station Journal Series No. 18-913077.

CHAPTER 8

SYMPTOMLESS GRASS ENDOPHYTES: PRODUCTS OF COEVOLUTIONARY SYMBIOSES AND THEIR ROLE IN THE ECOLOGICAL ADAPTATIONS OF GRASSES

C. W. Bacon and N. S. Hill

Toxicology and Mycotoxin Research Unit,
Russell Research Center
USDA/ARS
Athens, GA 30613
and
Department of Crop and Soil Sciences,
University of Georgia
Athens, GA 30602

There is paleontological evidence for the existence of fungi long before the development of Angiosperms. Fossil records indicate that the association of fungi with plants dates back to the Silurian and Devonian periods, a time span of approximately 395 million years ago. Although some of these associations undoubtedly represent strictly parasitic relationships, others were casual and suggests mycorrhizal associations. One of the earliest records of a true endophyte suggesting a mutualistic symbiosis was found in the roots of the fossil tree *Amyelon radicans* from the Paleozoic Era (71). This is considered important not from the standpoint of the origin of endophytic symbioses, but from the point of view that plant-fungus associations entered close relationships very early in their evolution which persisted and resulted in very specific relationships, one of which we refer to as a mutualism. Mutualisms are interactions between individual organisms in which the realized or potential genetic fitness of each participant is increased by the actions of the other. The finding of this specific association in the early fossil records is considered of consequence because it indicates the importance of mutualism as a survival strategy. Indeed, mutualisms are considered important in the early colonization of land by plants (50).

Many species of grasses (Gramineae) of the subfamily Pooideae (Festucoidae) are associated with intercellular fungi (101,103). These fungi grow endophytically within seeds, leaves, culms, rhizomes, and meristems of grasses and never show external signs of infection or symptoms of a disease. The fungi associated with these grasses belong to the anamorphic genus *Acremonium* Link:Fr. section *Albo-lanosa* Morgan-Jones & Gams and consist primarily of two species, *A. coenophialum* and *A. lolii* Latch, Christensen, & Samuels. Although other species with similar symptomless habits have been delineated (see FUNGAL MUTUALISTS below), these will not be discussed relative to mutualistic expression since we have no strong information on the nature of their relationship. The species *A. coenophialum* and *A. lolii* are perennially associated with tall fescue (*Festuca arundinacea* Schreb.) and perennial ryegrass (*Lolium perenne* L.), respectively, and are therefore considered obligate conjunctive associations (8) which is in contrast to the terminology used for endophytes of woody plants (18). In this discussion we follow the recommendation of Schardl *et al.* (87) in which the fungus-grass association is called a symbiotum (plural symbiota), replacing the traditional endophyte-infected grass terminology.

To understand the need for a mutualistic relationship to develop between *Acremonium* species and tall fescue and perennial ryegrass, the systematics and evolution of both the endophyte and plant species will be reviewed and salient evolutionary events discussed which may have occurred that resulted in the present day mutualism. Since there is debate on the taxonomy of clavicipitaceous endophytes, recent technological developments at the molecular level have been incorporated into this discussion to clarify these problems.

The available information indicates that these symbiota are associated with at least four specific classes of chemicals. Additional information will be discussed which indicates that these symbiota, especially tall fescue, have a greater drought tolerance to abiotic stresses than non-symbiotic grasses. The adeptness of the fungal mutualists to produce secondary metabolites, many of which deter predators and confer abiotic stress resistances, will be discussed and related to a cost:benefit model for the best estimate of the mutualists' contribution toward fitness during the coevolution of these symbiota.

The Symbiota

The tall fescue and perennial ryegrass symbiota are defensive mutualisms (23) because of the overall competitive benefits (42) which are due to specific characteristics such as enhanced drought tolerance (4,76), increased tillering and growth (44,57), and increased resistance to herbivory from mammals and insects (20,46). The components of these two symbiota are called mutualists. Unlike other grass symbiota (27), as well as other symbiotic associations, once a tall fescue or ryegrass symbiotum is established the fungal and grass mutualists do not live apart, even briefly. Therefore, these symbiota are described as obligate conjunctive associations. The symbiota are seed disseminated, their association with the grass begins after germination of infected seed where upon they infect developing seedlings, and symbiotic perennation is guaranteed by infecting the meristematic areas and all vegetative organs except the roots of the mature plant.

The Fungal Mutualists

Acremonium by definition reproduces solely by conidia and, therefore, belongs to the form-class Deuteromycetes. However, several species produce conidiophores typical of *Acremonium* species but are also capable of sexual reproduction, the structures of which indicates that they are Ascomycetes (70). The genus *Acremonium* consists of species of fungi which were previously identified as *Cephalosporium, Paecilomyces, Gliomastix, Sagrahamala,* and *Verticillium* because of their colorless mycelium and unicellular conidia. When first described by Link, the genus *Acremonium* was an artificial genus generated to accommodate those anamorphic hyphomycetes with teleomorphs belonging to the Eurotiales, Hypocreales, and Sordariales. Two tribes (Clavicipiteae and Balansiae) have been identified within the family Clavicipitaceae and are delineated on the nature and degree of association with the ovaries and vegetative parts of a host plant (27). Such a classification system has a distinct disadvantage in that it fails to recognize phylogenetic relationships among the species. Because the tribe Balansiae includes species of *Epichloë* Fr. (Tul.), taxa which produce the anamorphic *Acremonium* state, e. g. *A. typhinum* (Fr.) Tul., it is assumed that the fungal endophyte of the genera *Festuca* and *Lolium* also has affinity to this tribe.

Onions and Brady's (70) description of *Acremonium* states that they are "relatively slow-growing organisms, attaining a diameter of less than 25 mm in 10 days on malt extract or oatmeal agars at 20 C. Hyphae are thin-walled and hyaline. Typically, the erect conidiogenous cells are formed singly, but conidiophores with simple or verticillate branching occur in some species. Conidiogenous cells usually delimited by a basal septum, occasionally continuous with the hyphae from which they are formed, narrowing in shape gradually toward the tip."

When Bacon *et al.* (10) first associated the tall fescue endophyte with livestock toxicity, they suggested it was the anamorph of *E. typhina*, not only because of identical hyphae when grown in their respective hosts and identical conidia when grown on cornmeal malt agar (CMM), but also because of Sampson's (85) earlier assignment. However, the endophyte of tall fescue had a slower growth rate on CMM than *E. typhina* isolated from bentgrass (*Agrostis perennans* L.). Earlier work identifying endophytes in tall fescue and chewing fescue (*Festuca rubra* L.) also identified them as the anamorphs of *E. typhina* (69,85). Not accepting that anamorphs of the same organism could differ in growth rates, Morgan-Jones and Gams (67) re-examined growth characteristics and morphology of the tall fescue endophyte and the anamorph of *E. typhina* isolated from orchardgrass (*Dactylis glomerata* L.). When grown on 2% malt extract agar the tall fescue endophyte grew more slowly than the *Epichloë* anamorph, but both fungi produced similar conidia. However, they found that the conidiogenous cell was nonseptate in the isolate from tall fescue, but septate at the base in the *Epichloë* anamorph. Both endophytes had solitary phialides, which prompted them to form the new section *Albo-lanosa* within *Acremonium* which allowed for the distinction between anamorphs of the Clavicipitaceae and other Deuteromycetes with verticillate conidiophores. Thus, the name *A. coenophialum* Morgan-Jones and Gams was given to the tall fescue endophyte and *A. typhinum* Morgan-Jones and Gams to the anamorph of *E. typhina* isolated from orchardgrass.

Latch *et al.* (57) isolated and grew *A. coenophialum* from tall fescue on corn meal dextrose agar and found a basal septum in conidiophores suggesting that, other than growth rate, *E. typhina*, *A. coenophialum*, and *A. lolii* were morphologically similar but distinctly different from other *Acremonium* species.

The lack of discernible morphological structures, which are consistent within each *Acremonium* isolate, is counter to definitive classification and suggests that molecular genetic analyses are necessary to assist in identifying the taxonomic relationships of the genus using the dual classification system (88). Molecular analysis has several advantages over subjective morphological analyses, a major advantage being that molecular sequences remain constant within a species regardless of the nutritional history of the organism. With fungi, it is not uncommon for an organism to have pleomorphic forms depending upon the medium upon which it is growing. Therefore, our taxonomic discussions of plants and endophytes have incorporated molecular data to clarify species relations where appropriate.

Both nuclear and cytoplasmic sources of DNA are amenable to determining phylogenetic differences among fungi (19,86), with nuclear DNA being a more conserved genetic system than that in the cytoplasm. Regardless of the system chosen, each nucleotide position represents a character from which point mutations can be observed (88). Use of restriction endonucleases can cleave specific sequences of DNA which have like sizes. This procedure can be used to isolate specific loci within a genome which can later be characterized for base sequences. Once base sequences are determined, regression techniques can be used to determine similarity among organisms.

Schardl *et al.* (88) used ribosomal DNA to characterize nucleotide sequences among clavicipitaceous endophytes. In their study they evaluated isolates previously described as *E. typhina* from perennial ryegrass (57) and creeping bentgrass (11), *A. coenophialum* from two cultivars of tall fescue (67) and from *Poa autumnalis* (102), *A. typhinum* from *F. rubra* (67), *A.*

uncinatum Gams, Petrini & Schmidt from *F. pratensis* L. (35), *A. lolii* from *L. perenne* (57), and *Atkinsonella hypoxylon* (Peck) Diehl from *Danthonia spicatata* (L.) Beauv. (21). Their results suggested a remarkable similarity of ribosomal RNA among all *Acremonium* organisms and that they were hybrids of and phylogenetically related to, and evolved from *E. typhina* (88). They also found that fungal mutualists did not necessarily coevolve with their grass hosts. In fact, as much dissimilarity occurred among *A. coenophialum* isolates from *F. arundinacea* as existed between *A. coenophialum* and *A. lolii* or *A. typhinum*. These findings raise doubts as to the true taxonomic derivation among species of *Acremonium* within sect. *Albo-lanosa*.

Interestingly, the endophyte *A. uncinatum* isolated from *F. pratensis*, a possible progenitor of *F. arundinacea* (see below), was closely related to *A. coenophialum* isolated from *F. arundinacea* but could not be considered the same (88). This suggests that *A. coenophialum* evolved from a different species of *Epichloë* than did *A. uncinatum* and raises the question as to whether *A. coenophialum* invaded *F. arundinacea* late in its evolutionary development or whether it is derived from yet another species of *Epichloë*.

The Grass Mutualists

As with fungi, plants historically have been placed into discrete taxon based primarily upon the morphological traits of their sexual tissues, but now includes leaf, stem, and root structure. Gould and Shaw (37) subdivided the Gramineae into six subfamilies based on morphological criteria, but further subdivided the subfamilies into tribes with the aid of cytogenetic information. They have placed perennial ryegrass and tall fescue into the Pooideae subfamily and the Poeae tribe because of their membranous ligules, multi-flowered spikelets, and basic chromosome number of 7. These two grasses are found growing in open areas of deciduous woodlands in temperate regions.

The genus *Lolium* L. is a native to Europe, temperate Asia, North Africa, and the North Atlantic island countries. It is a small genus which consists of eight species, all of which are diploid (93). All species are compatible with one another, readily forming hybrids naturally and artificially through plant breeding. The progeny of crosses have nondistinguishing variation for every taxonomic character (54). Three species, perennial ryegrass (*L. perenne*), annual ryegrass (*L. multiflorum* L.), and *L. rigidum* Guadin. are wind pollinated which results in the occurrence of hybrid individuals (93).

The genus *Festuca* L. includes approximately 80 species throughout the temperate regions of the world (94). Tall fescue (*F. arundinacea*) has a wide native distribution in temperate and cool climates throughout Europe, North Africa, and west and central Asia. Species within the genus range from diploids to decaploids (2N=70). Two species, tall fescue (*F. arundinacea* Schreb.) and meadow fescue (*F. pratensis* [syn. *elatior*]), are extremely similar morphologically but diverse cytogenetically (16,93).

Meadow fescue is a diploid plant while tall fescue is a hexaploid (2N=42). They readily cross with complete chromosome pairing of the *F. pratensis* chromosomes with one genome of *F. arundinacea*, suggesting that meadow fescue may be one progenitor of tall fescue. A tetraploid (2N=28) race of *Festuca* was found in France and Spain, which was identified as *F. arundinacea* var. *glaucescens* Boiss and is considered to be the other progenitor of tall fescue. This suggests that either or both progenitors could serve as a vector of endophyte for the hexaploid tall fescue as we know it today (16,93).

Defensive Chemicals and Mechanisms-
The Reason For Being

Anti-Mammalian Herbivory

Historically tall fescue and perennial ryegrass symbiota were associated with animal performance problems and toxicities. The most common of these are fescue toxicosis and ryegrass staggers (6,9,33). Plants with deterrents to ruminant herbivory should have a competitive edge over plants without such a mechanism. The groups of chemicals associated with anti-mammalian herbivory are ergot alkaloid and tremorgenic neurotoxins. These compounds have been shown to be either produced exclusively by the fungus as in the case of the tall fescue mutualist (63,73,108) or only in infected grasses in the case of perennial ryegrass (29,34,81). However, the ryegrass symbiotum also contains ergot alkaloids (82). This indicates a striking biochemical similarity between these two fungal mutualists and species of the ascomycetous tribe Balansieae of the family Clavicipitaceae, which all have the ability to produce ergot alkaloids (7,90). The ability to produce ergot alkaloids, as well as several taxonomic (11,27,84) and biochemical characteristics (87,88), have all indicated an affinity of *Acremonium* species with members of the Clavicipitaceae. This is discussed in more detail below.

The ergot alkaloids found in symbiotic grasses consist of both the ergopeptine and clavine types (63,107,108). These two groups differ from each other in presence of a peptide bond and attached amino acids in the former and a lack of this bond and amino acids in the latter. The most predominant is the ergopeptine alkaloids of which ergovaline is the major alkaloid of this group. These alkaloids vary in concentration from 0.01 to 3.0 µmg/g of plant (dry weight). Mammalian toxicity resulting from consuming toxic forages and specific ergot alkaloids are reported in earlier reviews (9,92,106).

The tremorgenic neurotoxins, commonly called the lolitrems, have been isolated only from the perennial ryegrass symbiotum. This class of compounds are considered responsible for ryegrass staggers of sheep (33,34), and consists of four biologically active compounds, all containing a complex indole isoprenoid ring system (33). The major lolitrem is lolitrem B and is found in perennial ryegrass within the range of 3 to 25 μg/g dry weight (90). This class of compounds, unlike the ergot alkaloids, have not been isolated from cultures of the fungus *A. lolii*, but an indole isoprenoid precursor, paxilline, is synthesized by the fungus in culture (96). In addition to finding paxilline in fungus cultures, it has also been reported from ryegrass seed (96). It is unknown if this compound is made by the fungus and converted to the lolitrems by the plant, or if under culture or other conditions there is an incomplete synthesis by the fungus.

Anti-Insect and Other Herbivory

The effects of the ryegrass and tall fescue symbiotia on insects are very complex (12,36,90,97) and specific (41,90), but nevertheless positive enough to warrant consideration as competitive factors within grass communities. The list of insects reported as being deterred or poisoned by these two grasses include several species of aphids (46,55,90), sod webworms (32), leafhoppers (53), Chinch bugs (31), crickets (2,5), corn flea beetle (53), black beetle (91), bluegrass billbug (3), flour beetles (23), and Argentine stem weevil (74,75). Infected grasses are also toxic to the lepidopteran larvae of fall armyworms (40) and species of *Crambus* (24). In addition to insects, there are reports of the infection status of grasses on decreased reproduction (51,98) and feeding (72) of several species of nematodes, as well as an interaction of mycorrhizal infection on the ability of infected perennial ryegrass to deter nematodes (12).

Peramine is the chemical considered responsible for the deterring activity of several insects, specifically the Argentine stem weevil (81). Peramine is a very simple alkaloid which has been reported in symbiotic perennial ryegrass and tall fescue (80,90). It is not known if the fungus makes this compound or if its synthesis requires a combination of fungus and grass. In addition to peramine, the loline alkaloids (N-formyl and N-acetyl loline) have also been implicated in the toxicity and deterring activity of one species of aphid, *Rhopalosiphum padi*, but affected another species only if the infected grass contained a number of other alkaloids (lolines, peramine, and/or lolitrems) (90). Several environmental factors may also be responsible for the final effect (61,65). The loline alkaloids are pyrrolizidine bases that are found in the tall fescue symbiotum in concentrations as high as 0.8% of the dry weight of tall fescue plants. These alkaloids have not been isolated from cultures of the fungus, and the occurrence of similar compounds in other higher plants (65) implies they may be products of the plant responding to infection from the endophyte. This implication is strengthened by the report of Belesky *et al.* (14), which indicates that the concentration of loline alkaloids reflects the extent of infection within the population.

The variety of defensive compounds found within symbiotic individuals in a population may have the greatest effect on deterrences to mammalian and insect herbivory. Long-term herbivory will lead to sub-populations, each chemically defined and based on its specific mixture of deterring compounds. In areas where there are no pressures from herbivory, there should be a mixture of ecotypes, including individuals totally devoid of insect resistances. Therefore, it is important when experimental studies of specific resistant mechanisms of symbiota are being conducted to use only appropriate individuals.

Abiotic Stress Tolerances

Experimental evidences for the occurrence of stress mechanisms in the tall fescue and perennial ryegrass symbiota are an outgrowth of the initial observation of Read and Camp (76) that two of three populations of symbiotic tall fescue were more drought resistant than tall fescue plants with a low level of infection. The use of the term stress follows a broad definition (8) and is intended to include those constitutive and exogenous constraints which limit vegetative and reproductive efforts of the grass within the association. Inherent in this definition is the important contribution made by the fungus, while that of the grass is ignored. This is a working definition and is used at the experimental level only to distinguish between benefits derived from fungi within the symbiota.

The basic survival potential of a plant from drought is determined by its mechanism used to survive a declining water supply (47). The tall fescue symbiotum is considered to tolerate drought primarily by turgor maintenance (97), although stomatal responses have been implicated (13). In the event turgor maintenance is the major mechanism, the osmotic substance(s) involved should not be metabolizable, particularly during the period of drought. Each mutualist may contribute to the process of solute accumulation.

While the mechanisms for drought tolerance to water deficits in the tall fescue symbiotum are unknown, the ability of symbiotic grasses to develop low osmotic potential may be of significance. West *et al.* (97) determined that only specific tissue types, young meristematic and elongating leaf, from the tall fescue symbiotum were capable of developing low osmotic potential in response to water stress.

In a subsequent experiment it was determined that tissue from the leaf sheath and blade of this symbiotum developed lower osmotic adjustments than the uninfected plants (28). It was concluded that since osmotic adjustments persisted in young and immature leaf blades, osmotic adjustments may be the mechanisms of persistence and tiller survival in the tall fescue symbiotum under intermittent drought (97,99). The distribution of this mechanism throughout the tall fescue symbiotum is unknown, but as found for the occurrence of the herbivore toxins, it is expected to vary. One variation has been reported by White (104) who studied the expression of osmotic adjustment in two clones of the tall fescue symbiotum. He concluded that cell wall elasticity appeared to explain the differences in turgor maintenance among the two clones (104).

The existence of any grass or fungus mediated regulatory mechanism responsible for the accumulation of an osmotic substance is unknown. Further, the nature of osmotica responsible for the increased efficiency within the symbiota is unknown, but polyols, sugars, and amino acids are considered as likely candidates (66). The essential substance or substances must not only be osmotically active, but also compatible (nontoxic) with the normal plant physiological processes. While not essential, it is also highly desirable that the substance be nonmetabolizable, at least during its period of usefulness. Polyols are normal metabolites of fungi, and it has been determined that one genotype of the tall fescue symbiotum contains two polyols, mannitol and arabitol (78). Further, arabitol accumulated in infected grasses only under drought stress (-1 MPa). While the concentration of polyols was not high enough to affect the overall osmotic pool in this genotype, the combined concentration of fructose and glucose was considered significant enough to do so (78).

Symbiota appear to have some advantage over nonsymbiotic grasses in the area of nitrogen utilization. The symbiota, again primarily tall fescue, show two expressions from nitrogen stress. One, the symbiota can utilize nitrogen much more efficiently than nonsymbiotic tall fescue. Two, in the presence of high levels of soil nitrogen, one class of toxins, the ergot alkaloids, is increased. In both cases more positive benefits are derived from the symbiota than the comparable uninfected grass.

The efficient utilization of low soil nitrogen by the tall fescue symbiota was reported by Arechavaleta *et al.* (4). The amounts of dry matter produced by the low level of nitrogen (11 mg/pot) was the same as the amount of dry matter produced by uninfected tall fescue at the high (220 mg/pot), medium (73 mg/pot), and low (11 mg/pot) levels of nitrogen (4). Of the many enzymes responsible for nitrogen utilization, glutamine synthetase is primarily responsible for the efficient utilization of nitrogen. When the activity of this enzyme within the tall fescue symbiotum was compared to the activity in the uninfected plants grown under low nitrogen, it was discovered that glutamine synthetase was higher in the symbiotum. The high level of this enzyme in symbiota grown under low soil levels was considered an efficient means of utilizing nitrogen (62). The ability to efficiently utilize low nitrogen apparently is not present in seedlings of symbiota, but develops in mature symbiota (22).

High levels of soil nitrogen increase the amount of dry matter produced by plants. However, high levels of nitrogen not only increase the amount of dry matter produced in the tall fescue symbiotum (4,26,44), but also the amount of the ergot alkaloid toxins (4,44,63). Thus, although there is an increase in growth from nitrogen, an increase amount of toxins in the foliage should reduce herbivory. The effects of soil nitrogen on the accumulation pattern of other herbivore toxins are unknown, but since they are also nitrogen containing compounds they may also be affected by high soil nitrogen.

Fungus toxins might also interact with low rates of nitrogen to deter pests. Mattson (64) determined that the total nitrogen content of plants affects the degree of insect herbivory. Based on his data, insect feeding on the grasses grown in low soil nitrogen would consume proportionately more leaf area of grass than insects feeding on high nitrogen fertilized grass. Lyons *et al.* (62) determined that the total free nitrogen concentration of leaves of tall fescue was significantly decreased by endophyte infection although the plants were fertilized with high rates of nitrogen (10 mM total nitrogen/pot). Under high nitrogen, increased herbivory is prevented due to the action of increased amounts of toxins, and within a mixed population of grasses, herbivory of symbiota would be less than herbivory of uninfected grasses. Thus, under a wide range of soil nitrogen levels, the toxins and/or nitrogen content of symbiotic genotypes may be utilized to offset any tendency to overgraze infected grasses.

Cost:Benefit Analysis of Mutualism

An ideal analysis of a mutualism should consider its polymorphic nature as expressed within a population (49). By this it is meant that the relationship must contain: 1) successful symbionts which contribute to the fitness of one another, i.e. mutualists, 2) unsuccessful symbionts which inhabit but contribute nothing to the fitness of one another, and 3) non-symbionts which make no attempt at cohabitation and therefore receive no benefit. Tall fescue and perennial ryegrass symbiota have a preponderance of cohabitating individuals that fall in the first category which will be analyzed below. Within the tall fescue symbiotum, there is evidence that non-mutualists exist (Hill, unpublished data), but it is not clear whether they appear as non-mutualists because of our incapacity to test them in the appropriate environment. Non-mutualists can be found among plants of each species, but in all certainty these plantings are 'agricultural artifacts' resulting from agronomic uses and practices. Our concern here is for the natural tendency of mutualists to remain and be perpetuated as symbionts. Free-living fungal components of the tall fescue and perennial ryegrass have yet to be found.

For a population to be mutualistic, the fitness of successful mutualists must be greater than unsuccessful or non-mutualists (105), and the total fitness of mutualists must be greater than the non-mutualists. Being obligatory, the fungal non-mutualist can not be used to estimate cost:benefit of the cohabitating organisms. Comparisons between successful mutualistic and non-mutualistic plants can be made by removing the endophyte from within the plant, thus comparing genetically identical plants with and without the endophyte. From these kinds of experiments, data can be fitted into the cost:benefit model proposed by Keeler (49) using the formula:

$$W_{mf} = g_o + p(b_m - c_m + s_m + f_m p), \quad \text{Eq (1)}.$$

Where:

W_{mf} is the weight of the mutualists growing together (symbiotic plant);
g_o is the weight of the non-mutualist (endophyte free plant);
p is the probability that the endophyte will encounter a suitable host;
b_m is the benefit received from the interaction;
c_m is the cost of the mutualist's (endophyte) service;
s_m is the energetic changes due to changes in host structures necessary to house the inhabitant (endophyte); and
f_m represents a feedback loop on host abundance.

In controlled experiments we can manipulate these variables in order to estimate the unknowns. For example, because the endophyte is obligatory and seed disseminated, the chance that the endophyte will find a suitable host is 100% and therefore p is equal to 1. The cost of the mutualist's service, c_m, is difficult to assess, but Belesky *et al.* (13) found that maintenance respiration of symbiotic tall fescue plants was marginally more than nonsymbiotic plants. Antibiotic compounds provided to the plant are in minute quantities (44,90), and fungal mutualists are confined to leaf sheaths and aboveground meristems (11,69), suggesting that c_m is essentially nil.

Because the *Acremonium* mutualists are symptomless, the energy expenditure on structural changes of the host (s_m) is zero, and the fungus provides no known nutrition to the plant. Benefits other than nutrition, such as water scavenging occur, which have similar effects but cannot be given the same numerical score. Growth regulators (25), and osmotic adjustments (45,104) occur, which enable the plant to better tolerate defoliation and drought when infected with endophytes (15). Symbiotic tall fescue has the ability to increase its osmotic potential (97) by increasing the sugar concentration (78), which not only improves the plant's ability to scavenge water, but also provides a readily available energy source for regrowth during ephemeral rainfall. Increased sugar concentration is not additional non-structural carbohydrate, but is a consequence of conversion of oligo- and polysaccharides to their monomeric forms. Therefore, Eq (1) can be simplified to:

$$W_{mf} = g_o + b_m. \quad \text{Eq (2)}$$

Thus, the comparison between genetically identical symbiotic and nonsymbiotic plants or homologous populations is the best estimate of the endophyte's contribution towards fitness of the plant, Eq (2). The remaining discussion in this section is based on data which made comparisons between homologous populations or genetically identical plants which were either infected or not infected with the endophyte.

Under field conditions, symbiotic populations persist longer and compete better than nonsymbiotic populations of tall fescue and perennial ryegrass (22,75,98). As discussed earlier, components of improved fitness of symbiotic populations appear to be multifaceted, including improved vigor (44,57), changes in morphology (44), resistance to grazing (76), insect resistance (23,75), nematode resistance (51,98), disease resistance (103,109), and drought tolerance (4,28). The chemical basis for each component of fitness is not completely understood, but data defining cause and effect relationships are eminent. Similar chemical compounds which are responsible for improved fitness in tall fescue are found in ryegrass, but few experiments have been reported documenting the benefits of endophytes in ryegrass populations other than providing insect resistance. Inasmuch as fungal mutualists of the two grass species are essentially identical and the grass plants belong to the same tribe of the Poaceae, it might be appropriate to use tall fescue data as the model for both plant species.

The effects of the endophyte upon its host plant are not constant among plant genotypes or environments. Mineral nutrition, especially nitrogen, is essential for production of alkaloids by the endophyte (63) and tillering capacity by the plant (4). Undoubtedly, nutritional effects on plants would affect nutritional supply to the endophyte which may not make these findings surprising, but the presence of endophytes increases glutamine synthetase activity in plants (62).

Characteristics which are associated with drought tolerance are not expressed unless the plant is exposed to prolonged drought conditions. Roots of symbiotic plants will grow faster and deeper into a soil profile than genetically identical nonsymbiotic plants under drought conditions, but do not differ when grown at field capacity (79). Total reserve carbohydrate content and structures are similar among plants regardless of endophyte content when soil water content is at field capacity (-0.1 bars), but sugar monomers increase and polymers decrease in symbiotic tall fescue once drought stress conditions are imposed (78). Plant morphological changes as a result of symbiosis are not constant among tall fescue genotypes (75). Tillering capacity, specific leaf weight, and crown weight may increase, decrease, or remain the same depending upon the genotype of the infected plant.

The degree of chemical expression by the fungal mutualist depends not only on the biochemical competence of the fungus but also upon the genotype of each plant (43). Further, the ergovaline content of the tall fescue symbiotum increases with increased leaf area while others remain unchanged. This raises the issue as to whether the controlling mechanism is associated with the plant, the endophyte, or an interaction between the two. By inserting endophytes into a common tall fescue genotype (48) and by conducting genetic studies between high- and low-ergovaline producing symbiotic plant genotypes (1), it has been documented that the plant can regulate the expression of ergovaline production by the endophyte (43).

These data provide overwhelming evidence that interactions between plant and fungus, as well as environment with symbiotic plants, increase the phenotypic variation among tall fescue plants. They also suggest that symbiotic plants do not have all mechanisms of fitness imparted upon them by the endophyte, but it is the sum of all the plants in the population which express the fitness characteristics. Therefore, symbiotic populations of grasses are far more plastic and adaptable than nonsymbiotic populations (89). How fast symbiotic plants will encroach upon nonsymbiotic plants in a mixed community will depend upon the types and severity of stresses imposed upon the plants in their environment. Undoubtedly, severe or repeated insect grazing will eliminate nonsymbiotic tall fescue plants, while opportunistic insects will have minimal effect on the competitiveness of a nonsymbiotic plant. The physical environment of the plant (i.e. soil type, cation exchange capacity and water holding capacity of the soil) has a major impact on competitiveness of the nonsymbiotic plant (42). Generally, the more ideal the growing conditions, the better the ability of the nonsymbiotic plant to compete with its symbiotic counterpart.

Coevolution of Ryegrass and Tall Fescue with Their Endophytes

The evolutionary tendencies in the Balansiae as expressed by morphology (27,56,59,83,84), host distribution (27,101,103), the degree of parasitism and the change to mutualism (8,22,23,28,42), the change from the epibiotic habit to the obligate endophytic habit with the loss of sexuality (27,85,100,102), molecular genetics (87,88), and the presence of specific secondary metabolites (33,63,73,81,82,90,108) suggest that the *Acremonium* mutualists are related to and probably represent the climatic taxon within the phylogeny of the Balansiae (Fig. 1). The predominantly monophyletic viewpoint expressed within taxa in Fig. 1, is presented as such for the sake of simplicity. Indeed, the molecular data (87,88) suggest that the *Acremonium* species, and possible other endophytes, arose on numerous occasions, indicative ultimately of a polyphyletic origin.

The phylogeny of the clavicipitaceous fungi as expressed in Fig. 1 indicates that there are two major groups within this family, one subfamily is parasitic on insects (Cordycepitoideae typified by species of *Cordyceps* (Fr.) Link), the other subfamily, Clavicipitoideae, is parasitic on grasses and sedges. Within the Clavicipitoideae, four major tendencies are observed: 1) a line indicating localized parasites of grass ovaries as shown

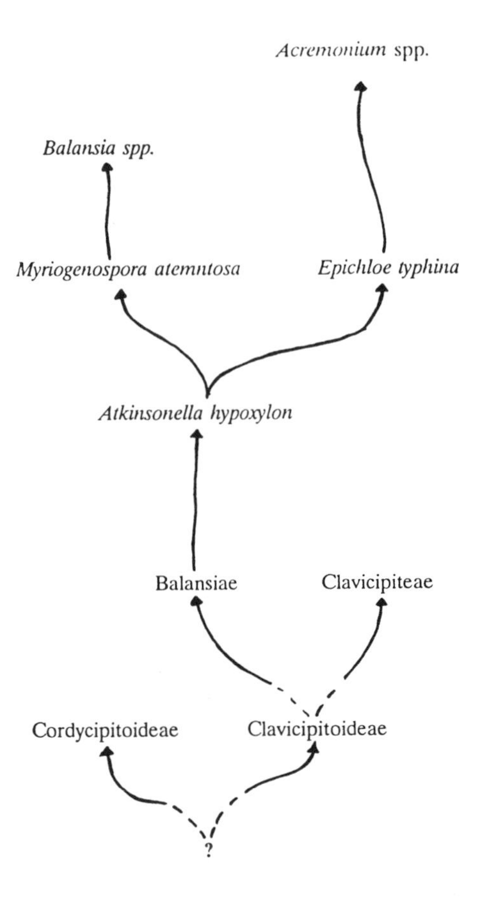

Fig. 1. Possible phylogeny in the family Clavicipitaceae based primarily on comparative morphology of the fungi and grasses. The question mark represents the unknown primitive Ascomycete from which the two subfamilies Clavicipitoideae and Cordycepitoideae originated.

by the *Claviceps* Tul. (tribe Clavicipiteae) and also characterized by having a stage independent of the grass, the sclerotium; 2) the change from a strictly ovarian parasitic location to include an epibiotic habit with two to one conidial states [presently shown by species of *Atkinsonella, and Myriogenospora atramentosa* (Berk. & M. A. Curtis) Diehl]; 3) development of the epibiotic habit to include an expanded endophytic habit, although somewhat antagonistic to grasses, which culminates with species of *Balansia*; and 4) *E. typhina* which is similar in degree of endophytic habit and antagonism as in the *Balansia* spp. but which culminates in the *Acremonium* spp. that are characterized by a restriction in the endophytic habit to leaf sheaths, loss of both the epibiotic habit and sexuality, no independent stage, seedborne dissemination, and the development of a strong mutualism. Thus, the placement of *A. hypoxylon* as one of the most primitive members of the Balansiae is based on the presence of both ephelidial (*Balansia*-type) and typhodial (*Acremonium-Epichloë*-type) conidia (54,84).

The ability of the two pathogenic organisms to invade viable seeds is mutualistically complementary of their inability to produce reproductive structures (58). If a species lives within a biotic environment of another mutualistic species (i.e., endophytes in their hosts), sexual recombination should operate to the benefit of the host because it is the organism which needs to respond to antagonistic environments. In the case of endophytes and their hosts, genetic variability of the host is essential so that the environment can select host phenotypes compatible with those growing conditions. The fungal mutualists (i.e., *Acremonium*) on the other hand, have other requirements for survival in their hosts and, having already achieved that end, receive no benefit from sexual reproduction. The algorithm has been proven using genetically identical tall fescue plants with or without their fungal mutualists. Regardless of host genotype, symbiotic grasses produced more seeds than their nonsymbiotic sister plants (42,77). Choke of orchardgrass caused by *E. typhina* (52) suggests that this symbiotum has not yet met criteria necessary for mutualism. A similar uncertainty exists for yet another orchardgrass symbiotum (*A. chilense-D. glomerata*) which is found growing in the mountains of Chile (68). This grass species which is associated with two different fungal species affords a unique tool for studies on the ecological success of a grass as contributed by two different fungal endophytes.

Acremonium is maternally transmitted to tall fescue and ryegrass seed (85,100); but it is unknown at what time during the evolution of the grass species that the transformation from pathogen to mutualist took place. The sexual compatibility of tall fescue with perennial ryegrass (17), combined with genetic similarity of their fungal mutualists (88), raise the question that evolution of mutualism with *Acremonium* endophytes occurred in a common plant ancestor. This would suggest that the phylogenetic branching of endophytes would be congruent with that of tall fescue and ryegrass (39). The fact that the endophyte of the tall fescue progenitor, meadow fescue (*F. pratensis* L.), is not as closely related to endophytes of either tall fescue or ryegrass as they are to one another (88), does not support this hypothesis. It certainly suggests that there are additional species of *Epichloë* and it is these that formed the basis of the symptomless *Acremonium* types. However, since there is only one known species, this suggestion is posed with hesitancy.

In addition to the hypothesis expressed earlier (88), the coevolution of *Acremonium* within perennial ryegrass occurred through natural outcrossing with other species of *Lolium*. In this scenario the endophyte could readily pass from one species to another through natural outcrossing and segregation of generations of progenies to their original parental forms. This suggests that nutritional environments for characteristics conducive to conjunctive endophytes would be fundamental in *Lolium* evolution. Sequencing of defined DNA segments and/or RFLP analysis of fungal mutualists within *Lolium* will be necessary to verify this hypothesis.

Coevolution of *Acremonium* with tall fescue is even more speculative. As already noted, the endophyte within one of the tall fescue progenitors is similar to that of *A. coenophialum* and *A. lolii* (88), but not to the extent that *A. coenophialum* and *A. lolii* are related to one another. However, this degree of unrelatedness may be reflective of endophyte changes and any speciation from the progenitors. Although there is no sexual state for these endophytes, ample mechanisms exist, e.g. parasexuality, which would allow for these fungi to become genetically diverse. It is possible that the tall fescue progenitor, *F. arundinacea* var. *glaucescens*, may contain a similar endophyte as *A. coenophialum*. However, natural occurrence of *F. arundinacea* var. *glaucescens* appears to be limited to a habitat in the mountainous regions of France and Spain and not widely adapted as is tall or meadow fescue (94). While its limited habitat does not necessarily mean the endophyte present in *F. arundinacea* var. *glaucescens* is not similar to that in tall fescue, it suggests that the endophyte which inhabits this progenitor does not confer similar benefits to allow its host to exploit diverse habitats as does the one in tall fescue. Again, a survey of *F. arundinacea* var. *glaucescens* is necessary to document that it contains an endophyte, and that its DNA sequence analyzed before its phylogeny can be determined with that of endophytes found in tall fescue. Clearly it is arguable but not documented if and when a common prehistoric ancestor existed to serve as a vehicle for endophyte dissemination through a segregating plant population.

Regardless of the evolutionary events which resulted in their cohabitation, it is clear that the *Acremonium* fungal mutualists meet the criterion that Law (58) proposed for mutualism. Those are: 1) inhabitants are genetically similar while their hosts are genetically diverse, 2) inhabitants rarely or never undergo sexual reproduction, and 3) inhabitants lack strong specificity for a particular host species.

Summary

All grasses are of a rather recent geological origin. The oldest reported fossil *Festuca* species was reported from the Miocene Epoch of the Tertiary period (95). Thus, we are concerned with a relatively young mutualism which, depending upon the time of establishing the association, must have developed within a time period of 26 million years ago.

The products considered responsible for extending the adaptability of grass symbiotia are secondary metabolites, several of which are produced by the fungal mutualist. Thus, the selective pressures from animal herbivory must have been important during the early period of grass evolution. As studies proceed we may find that the majority of the biologically active secondary metabolites are fungal products. Within the plant kingdom only a few groups lack the ability to produce of secondary metabolites, i.e., members of the Gramineae (110). Thus, the selective principles operating within these symbiota are probably based on secondary metabolism contributed by the fungal mutualists. Microorganisms that are rated as highly adaptable are also rated as having high biochemical abilities to utilize a wide variety of substrates and produce precursors from intermediary metabolism for use in primary metabolism (38). Such organisms are also characteristically slow growing, with long generation times (38). After isolation, the fungal mutualist of ryegrass and tall fescue grow slowly initially with several isolates being able to produce at least one class of secondary metabolites, the ergot alkaloids (7). Upon subculture on rich media their growth rate increases (personal observations of C. W. Bacon) along with a decrease in the qualitative and quantitative expression of ergot alkaloids (6). The fungi are intercellularly located, and nutrients are derived from the apoplasm which is nutritionally poor but contain a wide variety of compounds. The scarcity of nutrients within the apoplasm tends to direct the fungus toward continued production of secondary alkaloids, while its inability to reproduce sexually prevents it from deviating from this ability.

According to the symbiont theory of evolution (30) whole organisms such as bacteria or blue green algae were incorporated into early eukaryotic cells. The genetic information and its subsequent metabolism which were useful to the eukaryote were incorporated into its mitochondria and chloroplasts. We have no evidence of genetic information being incorporated into the grass chromosomes in the case of the fungal mutualists. The fungus is inserted within the grass, and the products of coevolution are a series of diverse secondary metabolites which played a part in the fitness of these grasses during their evolution.

Literature Cited

1. Agee, C. S., and Hill, N. S. 1992. Variability in progeny from high- and low-ergovaline producing tall fescue parents. Agron. Abstracts 83:185.
2. Ahmad, S., Govindarajan, S., Funk, C. R., and Johnson-Cicalese, J. M. 1985. Fatality of house crickets on perennial ryegrass infected with a fungal endophyte. Entomol. Exp. Appl. 39:183-190.
3. Ahmad, S., Johnson-Cicalese, J. M., Dickson, W.K., and Funk, C. R. 1986. Endophyte-enhanced resistance in perennial ryegrass to the bluegrass bill-bug *Sphenophorus parvalus*. Entomol. Exp. Appl. 41:3-10.
4. Arechavaleta, M., Bacon, C. W., Hoveland, C. S., and Radcliffe, D. E. 1989. Effect of the tall fescue endophyte on plant response to environmental stress. Agron. J. 81: 83-90.

5. Asay, K. H., Minnick, T. R., Garner, G. B., and Harmon, B. W. 1975. Use of crickets in a bioassay of forage quality in tall fescue. Crop Sci. 15:585-588.
6. Bacon, C. W. 1988. Procedure for isolating the endophyte from tall fescue and screening isolates for ergot alkaloids. Appl. Environ. Microbiol. 54:2615-2618.
7. Bacon, C.W. 1989. Isolation, culture and maintenance of endophytic fungi of grasses. Pages 259-282 in: The Isolation and Screening of Microorganisms from Nature. D. P. Labeda, ed. McGraw-Hill, New York.
8. Bacon, C.W. 1993. Abiotic Stress Tolerances (Moisture, Nutrients) and Photosynthesis in Endophyte-Infected Tall Fescue. Pages 123-141 in: *Acremonium*/Grass Interactions. S. Quisenberry and R. Joost, eds. Elsevier Science Publishers, Amsterdam.
9. Bacon, C. W., Lyons, P. C., Porter, J. K., and Robbins, J. D. 1986. Ergot toxicity from endophyte-infected grasses: A review. Agron. J. 78:106-116.
10. Bacon, C. W., Porter, J. K., Robbins, J. D., and Luttrell, E. S. 1977. *Epichloë typhina* from toxic tall fescue grasses. Appl. Environ. Microbiol. 34:576-581.
11. Bacon, C. W., and Siegel, M. R. 1988. The endophyte of tall fescue. J. Prod. Agric. 1:45-55.
12. Barker, G. M. 1988. Mycorrhizal infection influences *Acremonium*-induced resistance to Argentine stem weevil in ryegrass. Pages 199-203 in: Proc. 40th N.Z.Weed and Pest Control.
13. Belesky, D. P., Devine, O. J., Pallas, J. E. Jr., and Stringer, W. C. 1987. Photosynthetic activity of tall fescue as influenced by a fungal endophyte. Photosynthetica 21:82-87.
14. Belesky, D. P., Robbins, J. D., Stuedemann, J. A., Wilkinson, S. R., and Devine, O. J. 1987. Fungal endophyte infection-loline derivative alkaloid concentration of grazed tall fescue. Agron. J. 79:217-220.
15. Belesky, D. P., Stringer, W. C., and Hill, N. S. 1989. Influence of endophyte and water regime upon tall fescue accessions. I. Growth Characteristics. Ann. Bot. 63:495-503.
16. Bradshaw, A. D., Chadwick, M. J., Jowett, D., and Snaydon, R. W. 1964. Experimental investigations into the mineral nutrition of several grass species. J. Ecol. 52:665-676.
17. Buckner, R. C., Bush, L. P., and Burrus, P. B. 1973. Variability and heritability of perloline in *Festuca* sp. and *Lolium-Festuca* hybrids. Crop Sci. 13:666-669.
18. Carroll, G. 1988. Fungal endophytes in stems and leaves: From latent pathogen to mutualistic symbiont. Ecology 69:2-9.
19. Caten, C.E. 1986. The genetic integration of fungal life styles. Evolutionary biology of the fungi. Cambridge University Press, MA.

20. Cheplick, G. P., and Clay, K. 1988. Acquired chemical defenses in grasses: The role of fungal endophytes. Oikos 52:309-318.
21. Clay, K. 1984. The effect of the fungus *Atkinsonella hypoxylon* (Clavicipitaceae) on the reproductive system and demography of the grass *Danthonia spicata*. New Phytol. 98:165-175.
22. Clay, K. 1987. Effects of fungal endophytes on the seed and seedling biology of *Lolium perenne* and *Festuca arundinacea*. Oecologia 73:358-362.
23. Clay, K. 1988. Fungal endophytes of grasses: A defensive mutualism between plants and fungi. Ecology 69:10-16.
24. Clay, K., Hardy, T. N., and Hammond., A. M., Jr. 1985. Fungal endophytes of grasses and their effects on an insect herbivore. Oecologia 66:1-6.
25. De Battista, J. P., Bacon, C. W., Severson, R., Plattner, R. D., and Bouton, J. H. 1990. Indole acetic acid production by the fungal endophyte of tall fescue. Agron. J. 82:878-880.
26. De Battista, J. P., Bouton, J. H., Bacon, C. W., and Siegel, M. R. 1990. Rhizome and herbage production of endophyte-removed tall fescue clones and populations. Agron. J. 82:651-654.
27. Diehl, W.W. 1950. *Balansia* and Balansiae in America. Pages 1-82 in: USDA Agric. Monograph 4. U.S. Govt. Print. Office, Washington, DC.
28. Elmi, A. A., and West, C. P. 1989. Endophyte effect on leaf osmotic adjustments in tall fescue. Agron. Abst. 81: 111.
29. Fletcher, L. R., and Harvey, I. C. 1982. An association of a *Lolium* endophyte with ryegrass staggers. N.Z. Vet. J. 29:185-186.
30. Frederick, J. F. 1981. Origins and evolution of eukaryotic intercellular organelles. N.Y. Acad. Sci.
31. Funk, C.R., Halisky, P.M., Ahmad, S., and Hurley, R.H. 1985. How endophytes modify turfgrass performance and response to insect pests in turfgrass breeding and evaluation trials. Pages 137-145 in: Proceedings 5th International Research Conference, Avignon. F. Lemaire, ed. Versailles: INRA.
32. Funk, C. R., Halisky, P. M., Johnson, M. C., Siegel, M. R., and Stewart, A. V. 1983. An endophytic fungus and resistance to sod webworms. Biotechnology 1:189-191.
33. Gallagher, R. T., Hawkes, A. D., Steyn, P. S., and Vleggaar, R. 1984. Tremorgenic neurotoxins from perennial ryegrass causing ryegrass staggers disorder of livestock: Structure and elucidation of lolitrem B. J. Chem. Soc., Chem Commun. 1984:614-616.
34. Gallagher, R. T., Smith, G. S., Di Menna, M. E., and Young, P. W. 1982. Some observations on neurotoxin production in perennial ryegrass. N.Z.Vet.J. 30:203-204.
35. Gams, W., Petrini, O., and Schimdt, D. 1990. *Acremonium uncinatum*, a new endophyte in *Festuca pratensis*. Mycotaxon 37:67-71.

36. Gaynor, D.L., and Hunt, W.F. 1983. The relationship between nitrogen supply, endophytic fungus, and Argentine stem weevil resistance in ryegrass. Proc. N.Z. Grassland Assoc. 44:257-263
37. Gould, F.W. and Shaw, R.B. 1983. Grass Systematics. Texas A&M University Press, College Station, TX.
38. Groger, D. 1978. Ergot alkaloids. Pages 201-217 in: Antibiotics and other secondary metabolites. R. Hutter, T. Leisinger, J. Nuesch, and W. Wehrli, eds. Academic Press, New York.
39. Hafner, M. S., and Nadler, S. A. 1988. Phylogenetic trees support the coevolution of parasites and their hosts. Nature (London) 332:258-259.
40. Hardy, T. N., Clay, K., and Hammond, A. M., Jr. 1985. Fall armyworm (Lepidoptera: Noctuidae): A laboratory bioassay and larval preference study for the fungal endophyte of perennial ryegrass J. Econ. Entomol. 78:571-575.
41. Hardy, T. N., Clay, K., and Hammond, A. M. Jr. 1986. Leaf age and related factors affecting endophyte-mediated resistance to fall armyworm (Lepidoptera: Noctuidae) in tall fescue. Environ. Entomol. 15:1083-1089.
42. Hill, N. S., Belesky, D. P., and Stringer, W. C. 1991. Competitiveness of tall fescue as influenced by *Acremonium coenophialum*. Crop Sci. 31:185-190.
43. Hill, N. S., Parrott, W. A., and Pope, D. D. 1991. Ergopeptine alkaloid production by endophytes in a common tall fescue genotype. Crop Sci. 31:1545-1547.
44. Hill, N. S., Stringer, W. C., Rottinghaus, G. E., Belesky, D. P., Parrott, W. A., and Pope, D. D. 1990. Growth, morphological, and chemical component responses of tall fescue to *Acremonium coenophialum* Crop Sci. 30:156-161.
45. Izekor, E., West, C. P., Elmi, A. A., and Turner, K. E. 1989. Endophyte effect on tiller density and yield of water-stressed tall fescue. Agron. Abst. 81:135. (Abstr.)
46. Johnson, M. C., Dahlman, D. L., Siegel, M. R., Bush, L. P., and Latch, G. C. M. 1985. Insect feeding deterrents in endophyte-infected tall fescue. Appl. Environ. Microbiol. 49:568-571.
47. Jones, M.M., Turner, N.C., and Osmond, C.B. 1981. Mechanisms of drought resistance. Pages 15-37 in: Drought Resistance in Plants. L. G. Paleg, and D. Aspinall, eds. Academic Press, New York.
48. Kearney, J. F., Parrott, W. A., and Hill, N. S. 1991. Infection of somatic embryos of tall fescue with *Acremonium coenophialum*. Crop Sci. 31:979-984.
49. Keeler, K.H. 1985. Cost:benefit models of mutualism. Pages 100-127 in: The biology of mutualism:Ecology and evolution. Croom Helm, London.
50. Kelley, A.P. 1950. Mycotrophy in Plants. Chronica Botanica Co., Waltham, MA. 223 pp.

51. Kimmons, C. A., Gwinn, K. D., and Bernard, E. C. 1990. Nematode reproduction on endophyte-infected and endophyte-free tall fescue. Plant Dis. 74:757-761.
52. Kirby, E. J. M. 1961. Host-parasite relations in the choke disease of grasses. Trans. Br. Mycol. Soc. 44:493-503.
53. Kirfman, G. W., Brandenburg, R. L., and Garner, G. B. 1986. Relationship between insect abundance and endophyte infestation level in tall fescue in Missouri. J. Kans. Entomol. Soc. 59:552-554.
54. Kloot, P. M. 1983. The genus *Lolium* in Australia. Amer. J. Bot. 31:421-435.
55. Latch, G. C. M., Christensen, M. J., and Gaynor, D. L. 1985. Aphid detection of endophyte infection in tall fescue. N.Z.J. Agric. Res. 28:129-132.
56. Latch, G. C. M., Christensen, M. J., and Samuels, G. J. 1984. Five endophytes of *Lolium* and *Festuca* in New Zealand. Mycotaxon 20:535-550.
57. Latch, G. C. M., Hunt, W. F., and Musgrave, D. R. 1985. Endophytic fungi affect growth of perennial ryegrass. N.Z.J.Agric.Res. 28:165-168.
58. Law, R. 1985. Evolution in a mutualistic environment. Pages 145-170 in: The Biology of Mutualism: Ecology and Evolution. D. H. Bougher, ed. Croom Helm, London.
59. Luttrell, E. S. 1979. Host-parasite relationships and development of the ergot sclerotium in *Claviceps purpurea.* Can. J. Bot. 58:942-958.
60. Luttrell, E. S., and Bacon, C. W. 1977. Classification of *Myriogenospora* in the Clavicipitaceae. Can. J. Bot. 55:2090-2097.
61. Lyons, P. C. 1985. Infection and in vitro ergot alkaloid synthesis by the tall fescue endophyte and effects of the fungus on host nitrogen metabolism. Ph. D. Thesis. University of Georgia, College of Agriculture, Department of Plant Pathology, Athens. 124 pp.
62. Lyons, P. C., Evans, J. J., and Bacon, C. W. 1990. Effects of the fungal endophyte *Acremonium coenophialum* on nitrogen accumulation and metabolism in tall fescue. Plant Physiol. 92:726-732.
63. Lyons, P. C., Plattner, R. D., and Bacon, C. W. 1986. Occurrence of peptide and clavine ergot alkaloids in tall fescue. Science 232:487-489.
64. Mattson, W. S. 1980. Herbivory in relation to plant nitrogen content. Ann. Rev. Ecol. Syst. 11:119-161.
65. McLean, E. K. 1970. The toxic actions of pyrrolizidine (*Senecio*) alkaloids. Pharm. Rev. 22:429-483.
66. Morgan, J. M. 1984. Osmoregulation and water stress in higher plants. Ann. Rev. Pl. Physiol. 35:299-319.
67. Morgan-Jones, G., and Gams, W. 1982. Notes on Hyphomycetes, XLI. An Endophyte of *Festuca arundinacea* and the anamorph of *Epichloë typhina*, new taxa in one of two new sections of *Acremonium*. Mycotaxon 15:311-318.

68. Morgan-Jones, G., White, J. F.,Jr., and Piontelli, E. L. 1990. *Acremonium chilense*, and undescribed endophyte occurring in *Dactylis glomerata* in Chile. Mycotaxon 39:441-445.
69. Neill, J. C. 1941. The endophytes of *Lolium* and *Festuca*. N. Z. J. Sci. Technol. 23:185A-193A.
70. Onions, A.H.S., and Brady, B.L. 1987. Taxonomy of *Penicillium* and *Acremonium*. Pages 9-36. in: *Penicillium* and *Acremonium*. J. F. Peperdy, ed. Plenum Press, New York.
71. Osborn, T. G. B. 1909. The lateral roots of *Amyelon radicans* and their mycorrhiza. Ann. Bot. 23:603-611.
72. Pedersen, J. F., Rodriguez-Kabana, R., and Shelby, R. A. 1988. Ryegrass cultivars and endophyte in tall fescue affect nematodes in grass and succeeding soybean. Agron. J. 80:811-814.
73. Porter, J. K., Bacon, C. W., and Robbins, J. D. 1979. Ergosine, ergosinine, chanoclavine I from *Epichloë typhina*. J. Agric. Food Chem 27:595-598.
74. Pottinger, R.P., Barker, G.M., and Prestidge, R.A. 1985. A review of the relationships between endophytic fungi of grasses (*Acremonium* spp.) and Argentine stem weevil (*Listronotus bonarienses* Kuschel). in: Proceedings 4th Australasian Conference Grassland Invertebrate Ecology, 38: 322-331.
75. Prestidge, R.A., Pottinger, R.P., and Barker, G.M. 1982. An association of *Lolium* endophyte with ryegrass resistance to Argentine stem weevil. Proceedings New Zealand Weed Pest Control Conference. 35:119-122.
76. Read, J. C., and Camp, B. J. 1986. The effect of the fungal endophyte *Acremonium coenophialum* in tall fescue on animal performance, toxicity, and stand maintenance. Agron. J. 78:848-850.
77. Rice, J. S., Pinkerton, B. W., Stringer, W. C., and Undersander, D. J. 1990. Seed production in tall fescue as affected by fungal endophyte. Crop Sci. 30:1303-1305.
78. Richardson, M. D., Chapman, G. W., Hoveland, C. S., and Bacon, C. W. 1992. Sugar alcohols in endophyte-infected tall fescue under drought. Crop Sci. 32:1060-1061.
79. Richardson, M. D., Hill, N. S., and Hoveland, C. S. 1990. Rooting patterns of endophyte infected tall fescue grown under drought stress. Agron. Abst. 81:129.
80. Rowan, D. D., and Gaynor, D. L. 1986. Isolation of feeding deterrents against Argentine stem weevil from ryegrass infected with the endophyte. J. Chem. Ecol. 12:647-658.
81. Rowan, D. D., Hunt, M. B., and Gaynor, D. L. 1986. Peramine, a novel insect feeding deterrent from ryegrass infected with the endophyte *Acremonium loliae*. J. Chem. Soc., Chem Commun. 935-936.
82. Rowan, D. D., and Shaw, G. J. 1987. Detection of ergopeptine alkaloids in endophyte-infected perennial ryegrass by tandem mass spectrometry. N. Z. Vet. J. 35:197-198.

83. Rykard, D. M., Bacon, C. W., and Luttrell, E. S. 1985. Host relations of *Myriogenospora atramentosa* and *Balansia epichloë* (Clavicipitaceae). Phytopathology 75:950-956.
84. Rykard, D. M., Luttrell, E. S., and Bacon, C. W. 1984. Conidiogenesis and conidiomata in the Clavicipitoideae. Mycologia 76:1095-1103.
85. Sampson, K. 1933. The systemic infection of grasses by *Epichloë typhina* (Pers.) Tul. Trans. Br. Mycol. Soc. 18:30-47.
86. Scazzocchio, C. 1986. The natural history of fungal mitochondrial genomes. Pages 53-74 in: Evolutionary Biology of the Fungi. A. D. M. Rayner, C. M. Brasier, and D. Moor, eds. Cambridge University Press, Cambridge.
87. Schardl, C. L., Liu, J. S., White, J. F., Jr., Finkel, R. A., and Siegel, M. R. 1991. Molecular phylogenetic relationships of nonpathogenic grass mycosymbionts and clavicipitaceous plant pathogens. Pl. Syst.. Evol. 178:27-41.
88. Schardl, C.L., and Siegel, M.R. 1992. Molecular genetics of *Epichloë typhina* and *Acremonium coenophialum*. Agriculture Ecosystems and Environment. 44:169-185.
89. Schmid, B. 1985. Clonal growth in grassland perennials. III. Genetic variation and plasticity between and within populations of *Bellis perennis* and *Prunella vulgaris*. J. Ecol. 73:819-830.
90. Siegel, M. R., Latch, G. C. M., Bush, L. P., Fammin, N. F., Rowen, D. D., Tapper, B. A., Bacon, C. W., and Johnson, M. C. 1991. Alkaloids and insecticidal activity of grasses infected with fungal endophytes. J. Chem. Ecol. 16:3301-3315.
91. Siegel, M. R., Latch, G. C. M., and Johnson, M. C. 1987. Fungal endophyte of grasses. Ann. Rev. Phytopath. 25:293-315.
92. Stuedemann, J. A., and Hoveland, C. S. 1988. Fescue endophyte: History and impact on animal agriculture. J. Prod. Agric. 1:39-44.
93. Terrell, E. E. 1968. A Taxonomic Revision of the Genus *Lolium*. U.S. Dept. of Agric. Tech. Bull. No. 1392. U.S. Govt. Printing Office, Washington, DC.
94. Terrell, E. E. 1979. Taxonomy, morphology, and phylogeny. Pages 31-39. in: Tall Fescue. R. C. Buckner and L. P. Bush, eds. American Society of Agronomy, Madison, WI.
95. Thomasson, J.R. 1986. Fossil Grasses: 1820-1986 and Beyond. Pages 159-167 in: Grass Systematics and Evolution. T. R. Soderstrom, K. W. Hilu, C. S. Campbell and M. E. Barkworth, eds. Smithsonian Institution Press, Washington, DC.
96. Weedon, C. M., and Mantle, P. G. 1987. Paxilline biosynthesis by *Acremonium loliae*; a step towards defining the origin of lolitrem neurotoxins. Phytochemistry 26:969-971.
97. West, C.P., Izekor, E., Elmi, A., Robbins, R.T., and Turner, K.E. 1989. Endophyte effects on drought tolerance, nematode infestation and persistence of tall fescue. Pages 23-27 in: Proceeding of the Arkansas

Fescue Toxicosis Conference, No. 140. C. P. West, ed. Arkansas Agric. Exper. Sta., University of Arkansas, Fayetteville.

98. West, C. P., Izekor, E., Oosterhuis, D. M., and Robbins, R. T. 1990. The effect of *Acremonium coenophialum* on the growth and nematode infestation of tall fescue. Plant and Soil 112:3-6.
99. West, C. P., Oosterhuis, D. M., and Wullschleger, S. D. 1990. Osmotic adjustment in tissues of tall fescue in response to water deficit. Environ. Exp. Bot. 30:1-8.
100. Western, J. H., and Cavett, J. J. 1959. The choke disease of cocksfoot (*Dactylis glomerata*) caused by *Epichloë typhina* (Fr.) Tul. Trans. Br. Mycol. Soc. 42:298-307.
101. White, J. F., Jr. 1987. The widespread distribution of endophytes in the Poaceae. Plant Dis. 71:340-342.
102. White, J. F., Jr., and Bultman, T. L. 1987. Endophyte-host associations in forage grasses. VIII. Heterothallism in *Epichloë typhina*. Amer. J. Bot. 74:1716-1722.
103. White, J. F., Jr., and Cole, G. T. 1985. Endophyte-host associations in forage grasses. I. Distribution of fungal endophytes in some species of *Lolium* and *Festuca*. Mycologia 77:323-327.
104. White, R. H. 1989. Water relations characteristics of tall fescue as influenced by *Acremonium coenophialum*. Agron. Abst. 81:167.
105. Wilson, D. S. 1983. The effect of population structure on the evolution of mutualism: A field test involving burying beetles and their phoretic mites. Amer. Naturalist 121:851-870.
106. Yates, S.G. 1983. Tall fescue toxins. Pages 249-273 in: Handbook of Naturally Occurring Food Toxicants. M. Recheigel, ed. CRC Press, Boca Raton, FL.
107. Yates, S. G., Fenster, J. C., Bartelt, R. J., and Powell, R. G. 1987. Toxicity assay of tall fescue extracts, fractions and alkaloids using the large milkweed bug. Proc. Tall Fescue Mtg., Southern Region Information Exchange Group 37, Nov.17-18, Memphis, TN.
108. Yates, S. G., Plattner, R. D., and Garner, G. B. 1985. Detection of ergopeptine alkaloids in endophyte infected, toxic Ky-31 tall fescue by mass spectrometry/mass spectrometry. J. Agric. Food Chem 33:719-722.
109. Yoshihara, T., Togiya, S., Koshino, H., Sakamura, S., Shimanuki, T., Sato, T., and Tajimi, A. 1985. Three fungitoxic sesquiterpenes from stromata of *Epichloë typhina*. Tetra. Let. 26:5551-5554.
110. Zahner, H., Anke, H., and Anke, T. 1983. Evolution and secondary pathways. Pages 153-171 in: Secondary Metabolism and Differentiation in Fungi. J. W. Bennett and A. Ciegler, eds. Marcel Dekker, New York.

CHAPTER 9

RESOURCES AND TESTING OF ENDOPHYTE-INFECTED GERMPLASM IN NATIONAL GRASS REPOSITORY COLLECTIONS

A. D. Wilson

USDA Forest Service
Southern Hardwoods Laboratory
P.O. Box 227
Stoneville, MS 38776

Clavicipitaceous endophytes have been known to exist in grasses since the discovery of an endophyte in seeds of darnel (*Lolium temulentum* L.) by Vogl in 1898 (26). The oldest known specimens of darnel with endophytic mycelium were seeds retrieved from a pharoah's tomb in an Egyptian pyramid dating back to 3400 B.C. (16). Subsequent work by numerous investigators has shown that these fungi have hosts that are widely distributed in the grass family (Poaceae) and sporadically distributed in the sedge family (Cyperaceae) and rush family (Juncaceae) (3, 27). Most surveys for clavicipitaceous endophytes have concentrated on wild grasses and cultivated turfgrass and forage grass collections from Europe and North America (15, 20, 27). The extensive national collections of wild and cultivated grasses in United States repository collections have been largely ignored. The intent of this chapter is to increase general awareness of this large and valuable resource available to anyone doing research on grasses including endophyte researchers. Furthermore, this report summarizes initial grass endophyte research conducted by the author and colleagues since early 1989 on grass germplasm from these national repository collections.

Clavicipitaceous endophytes of grasses may be categorized into two groups: 1) the choke-inducing sexual forms or clavicipitaceous teleomorphic endophytes, and 2) the mutualistic asexual forms or clavicipitaceous anamorphic endophytes (29). Teleomorphic endophytes are ascomycetes (Clavicipitaceae: tribe Balansieae) that produce a sexual stage (teleomorph) on stromata and induce choke diseases which may disrupt reproduction of their hosts (8). Anamorphic endophytes are probably asexual derivatives of teleomorphic endophytes, but they lack a sexual stage, rarely sporulate in their hosts, and form mutualistic associations with their hosts. They are imperfect fungi (Deuteromycetes) classified in the genus *Acremonium* sect. *Albo-lanosa* and in related anamorphic genera (e.g. *Gliocladium* and *Phialophora*) (13, 17). Anamorphic endophytes infect primarily cool-season grasses of the tribe Pooideae (4). The endophytic fungi considered in this paper are primarily the anamorphic forms.

A number of general characteristics *in vivo* distinguish clavicipitaceous anamorphic endophytes, particularly the *Acremonium* sp., from other endophytic fungi of grasses. Morphologically, anamorphic endophytes tend to have relatively fine hyphae (≤4 μm diameter) that are often convoluted, highly vacuolated, infrequently branched, occasionally constricted at the septa, and have septa that stain poorly or not at all. The anamorphic forms are mutualistic, obligate endosymbionts (biotrophs) producing symptomless infections of their hosts, often attributed to the absence of teleomorphic states. Most species grow entirely in the vegetative (somatic) phase throughout their life cycles *in vivo*, although some species produce conidia in their hosts (18). However, most endophytes can be cultured and induced to sporulate on common growth media. The majority of anamorphic species are slow growing *in vitro* and are characterized by their white to tan, appressed and waxy to cottony colonies that often become wrinkled, puckering the agar in older cultures. Conidiophores arise singly, or form on synnemata or branched conidiophores. The fungi are seed borne and most commonly associated with the aleurone layer and adjacent tissues of grass seeds. They grow intercellularly and systemically through grass tillers mostly parallel to the longitudinal axis of the vascular system. These fungi overwinter as dormant inoculum in the meristematic region of their host. In addition, they produce ergot and related alkaloids throughout their hosts that probably contribute most to their biological activity. Finally, some species are also known to produce the growth-promoting phytohormone indoleacetic acid (auxin) in culture (7).

The infection of grasses by clavicipitaceous endophytes may have both beneficial and deleterious consequences with considerable economic significance. Endophytes can have profound effects on host physiology, reproductive biology, selection, ecology, growth, and pest resistance. The enhanced resistance in endophyte-infected grasses to many insects and diseases are most noted (2), but these fungi also have been implicated in increasing growth and vigor, water use efficiency, tolerance to heat and drought stress, competitive ability, photosynthetic efficiency and long-term survival of their hosts (23). Endophytes have been viewed as fortuitous sources of resistance in many grasses that lack genetic pest resistance. The turfgrass industries in the U.S. and abroad have taken advantage of these traits by releasing certified endophyte-infected cultivars with resistance to pests and environmental stresses. The selection advantages afforded by endophyte infection may significantly impact the ecology of graminaceous hosts through reduction of herbivory resulting from increased resistance to insects and ungulate, mammalian herbivores. However, the toxic effects of endophyte-infected grasses on livestock have caused large economic losses to the livestock industry. Toxicoses of livestock caused by the consumption of endophyte-infected grasses have resulted in annual losses of hundreds of millions of US dollars in lost production for both the United States and New Zealand (23).

National Plant Resources for Endophyte Research

The U.S. National Plant Germplasm System (NPGS), jointly administered by the Agricultural Research Service (ARS) of the United States Department of Agriculture (USDA) and the State Agricultural Experiment Stations (SAES), maintains the largest collections of herbaceous plants in the United States. Plant materials held in NPGS repositories represent a wide diversity of ecological habitats and plant genotypes from many countries throughout the world. The NPGS presently consists of 27 repository stations strategically located throughout the United States. Eleven plant repository stations contain grass collections rich in both genetic and endophyte-based resistance to agronomic pests. The 11 repository sites holding grasses include 4 regional plant introduction stations, 3 specialized or crop-specific collections, 2 national clonal repositories, and an importation quarantine office (Fig. 1). Most germplasm held in national repository collections is available to all researchers interested in testing this material for pest resistance or agronomic traits that may be useful for eventual introductions in crop improvement applications

The eleven national grass repository stations collectively maintain at least 235 grass genera representing over 1,600 species and more than 207,700 plant inventory (PI) accessions (Table 1). The major national cereal grass collections are located at only six repository sites (Table 2). The National Small Grains Collection (NSGC) in Aberdeen, Idaho holds the majority of the cereal grass collections including the wheat, barley, oats, rice, rye, and triticale collections. Some collections are divided between two or more sites for security. The NC-7, NSSL, and S-9 stations hold the major corn, sorghum, and millet collections. The CR-MIA station contains the sugarcane collection. Major national collections of wild grasses are less widely distributed at only two repository sites (Table 3) or at single repository sites (Table 4). Most national wild forage and turfgrass collections are located at the W-6 or Western Regional Plant Introduction Station (WRPIS) in Pullman, Washington and the National Seed Storage Laboratory (NSSL) in Fort Collins, Colorado. Other notable major wild grass collections with an abundance of PI accessions are available at the CR-MIA, NC-7, NE-9, NSGC, and S-9 stations (Table 4).

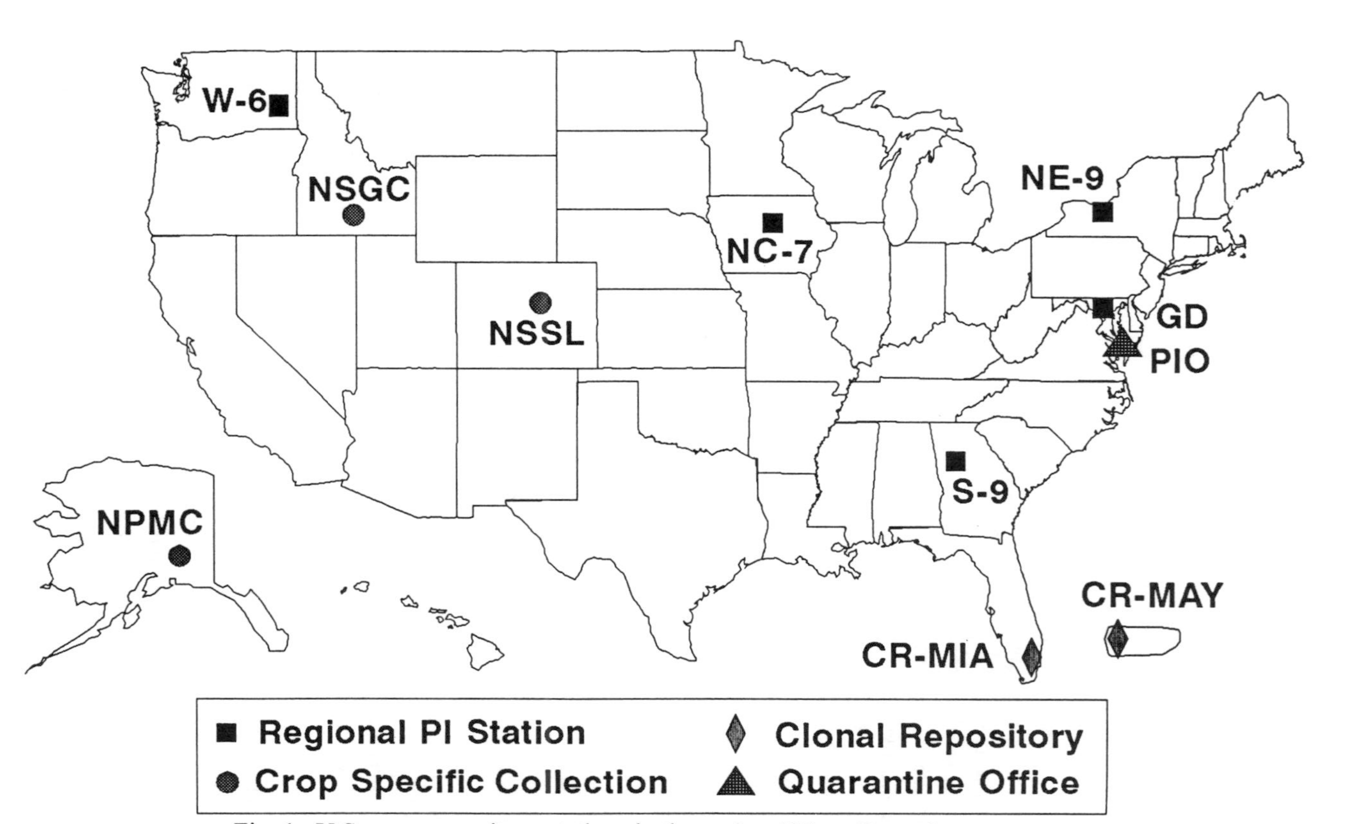

Fig. 1. U.S. grass repository stations in the national Plant Germplasm system.

Table 1. Summary of grass germplasm resources available from repository sites of the National Plant Germplasm System.[a]

Repository Sites and Locations	Station Code	Genera	Accessions
National Clonal Germplasm Repository (Mayaguez, Puerto Rico)	CR-MAY	17	97
National Clonal Germplasm Repository (Miami, Florida)	CR-MIA	19	2,510
Plant Germplasm Quarantine Office (Glenn Dale, Maryland)	GD	24	719
North Central Regional PI Station (Ames, Iowa)	NC-7	17	13,453
Northeast Regional PI Station (Geneva, New York)	NE-9	4	720
National Plant Materials Center (Palmer, Alaska)	NPMC	31	149
National Small Grains Collection (Aberdeen, Idaho)	NSGC	13	112,103
National Seed Storage Laboratory (Fort Collins, Colorado)	NSSL	62	40,019
Plant Introduction Office (Beltsville, Maryland)	PIO	32	465
Southern Regional PI Station (Griffin, Georgia)	S-9	107	25,505
Western Regional PI Station (Pullman, Washington)	W-6	74	11,984

[a]Germplasm available as of 20 May 1991. Data derived from selective searches of the Germplasm Resources Information Network (GRIN) data base.

All plant repository stations provide germplasm directly to users and research scientists worldwide. Information about specific collections is available to users through the NPGS electronic data base known as the Germplasm Resources Information Network (GRIN), a component of the Germplasm Services Laboratory at Beltsville, Maryland. The GRIN system is a public data base that is accessible to users by computer via modem and telephone links. Users can order materials electronically through the order-

Table 2. Major cereal grass collections in the National Plant Germplasm System.

Cultivated Collections	Representative Food Crops	National Plant Repository Stations[a]					
		CR-MIA	GD	NC-7	NSGC	NSSL	S-99S
Avena	Oats	-	-	-	22,513	630	-
Eleusine	African Millet	-	1	1	-	960	774
Hordeum	Barley	-	-	-	26,271	764	-
Oryza	Rice	-	615	-	16,108	745	-
Panicum	Common Millet	-	2	2,189	-	170	701
Pennisetum	Pearl Millet	1	3	-	-	764	638
Saccharum	Sugarcane	2,051	26	-	-	6	-
Secale	Rye	-	-	-	2,593	33	-
Sorghum	Sorghum	2	-	-	-	7,777	15,983
Triticum	Wheat	-	-	-	42,506	1,622	-
Triticosecale	Triticale	-	-	-	1,078	32	-
Zea	Corn	-	-	9,073	-	-	16,384

[a]Values indicate the number of accessions of each generic collection available as of 20 May 1991.

Table 3. Major wild grass collections distributed between two repository sites.

Wild Grass Collections	Common Names	Repository Stations[a]	
		NSSL	W-6
Agropyron	Wheatgrass	10	506
Agrostis	Bentgrass	14	213
Alopecuris	Foxtail	2	145
Bromus	Bromegrass	30	968
Dactylis	Orchardgrass	132	1,060
Elymus	Wildrye	10	1,131
Eragrostis	Lovegrass	14	1,295
Festuca	Fescue	162	1,503
Lolium	Ryegrass	117	844
Phalaris	Canarygrass	25	683
Poa	Bluegrass	93	667
Stipa	Needlegrass	1	271

[a]Values indicate the number of accessions of each generic collection available as of 20 May 1991.

Table 4. Major wild grass collections located at single Repository sites.

Wild Grass Collections	Common Names	Accessions Available	Repository Station
Aegilops	Goatgrass	1,014	NSGC
Andropogon	Beardgrass	1,066	NSSL
Bothriochloa	Bluestem	696	S-9
Cenchrus	Sandbur	826	S-9
Chloris	Fingergrass	246	S-9
Cynodon	Burmudagrass	505	S-9
Danthonia	Oatgrass	26	W-6
Digitaria	Crabgrass	661	S-9
Echinochloa	Hedgehoggrass	779	NC-7
Elytrigia	Wheatgrass	854	W-6
Erianthus	Plumegrass	209	CR-MIA
Holcus	Velvetgrass	18	W-6
Leymus	Dunegrass	346	W-6
Melica	Melicgrass	95	W-6
Oryzopsis	Ricegrass	231	W-6
Paspalum	Paspalum	1,492	S-9
Phleum	Timothy	567	NE-9
Setaria	Bristlegrass	1,325	NC-7
Sporobolus	Dropseed	118	W-6

[a]Values indicate the number of accessions of each generic collection available as of 20 May 1991.

processing system. Information of varying detail is available on approximately 370,000 items in the data base. Data on materials newly acquired from plant collectors are usually more complete.

Information in three categories may be obtained from the GRIN data base including passport, descriptors, and evaluation data (21). Passport data describe the origins and taxonomy of accessions. Descriptor data generated at the curator level provide information on the growth habit, ecology, and salient agronomic characteristics of each accession. Evaluation data are derived from performance evaluations and screening tests conducted by scientists which characterize the useful agronomic traits, pest resistance, genetics, and other unique attributes of particular accessions. For example, information on endophyte incidence in seeds of generic collections that have been surveyed can be found in the evaluation data on these collections. Data on PI accessions such as year original seed was received, number of increases since received, and year of last increase generally are not included in the GRIN data base, but such data are often available directly from curators. Curators of individual repository stations usually keep this information for the purpose of determining the need for and planning of future collection increases.

Certain materials have restricted availability for various reasons. Some germplasm has been patented which mandates by U.S. law that the recipient must contact the owner concerning restrictions on use. Inventories of materials requested regularly often become depleted to levels requiring that these accessions be temporarily removed from availability during the time when the material is being increased in the field or greenhouse. Other materials which are noxious weeds may have restrictions for shipment into certain regions.

Testing Grass Germplasm For Endophytes

Clavicipitaceous anamorphic endophytes are well known for their systemic, symptomless infections of grasses. Consequently, the incidence of clavicipitaceous endophytes in grass germplasm may be determined primarily by diligent microscopic examination or through chemical analyses to detect their alkaloid secondary metabolites. The economic significance and potential benefits of endophyte-based pest resistance as well as the hazards associated with livestock toxicoses have increased our need to determine their incidence in major world grass collections.

Surveys for clavicipitaceous endophytes in grasses of NPGS collections began in early 1989. The survey and characterization of anamorphic endophytes in the U.S. *Lolium* collection from the W-6 station marked the beginning of surveys of our national grass collections (29). Results of the U.S. *Lolium* collection survey have shown that anamorphic endophytes occur in wild ryegrasses in many countries of the world. Besides their commonly observed presence in *Lolium* germplasm from Europe, endophytes were found in seeds from Africa, Asia, Australia, and New Zealand. The highest incidence of infection occurred in seeds from the Middle East suggesting that this region may be a center of origin of clavicipitaceous endophyte-grass associations since many wild and cultivated grasses have originated there. The overall incidence of endophyte infection among surveyed accessions in the collection was relatively low (33%) when compared with many commercial grass collections. This trend was postulated to result from nonselective maintenance procedures used by U.S. germplasm repository personnel to avoid loss of potentially valuable traits.

Morphological studies of endophytic mycelia in seeds of five *Lolium* species revealed that considerable variation can occur in endophyte morphology in different host species (29). Despite the variation in morphological characteristics, there was still considerable overlap in endophyte morphologies across host species. Controlled experiments showed that the environmental conditions under which host plants are grown has considerable influence on hyphal diameters and the rates that endophytic hyphae of perennial ryegrass (*L. perenne* L.) enter leaf sheaths from seeds. Light intensity, soil fertility, and volume of growth media appeared to particularly influence endophyte growth and colonization of its host. The amount of mycelium in many PI accessions often was much lower than amounts observed in commercial cultivars (unpublished data). The low level of mycelium and infection rates in some accessions made detection difficult without relatively large sample sizes. In general, conditions favoring growth and vigor of the host also tended to support more vigorous growth of the endophyte. Furthermore, the systemic distribution of endophytic hyphae in host tissues depended on the particular host-endophyte association. For example, the endophytes of annual ryegrass species, unlike perennial ryegrass, poorly colonized leaf sheaths, but hyphae were found primarily in culm piths, especially in the internodes. Similar results were reported by Latch *et al.* (15).

The incidence of anamorphic endophytes in several other major grass collections including the fescue (*Festuca*), wild barley (*Hordeum*), and bluegrass (*Poa*) collections has been examined since the initial ryegrass survey (Table 5). The discovery of anamorphic endophytes in three wild *Hordeum* species demonstrated that these fungi also occur in the wild relatives of cereal grasses (30, 31, 32). The report of endophyte-infected accessions of *H. comosum* Presl. from Argentina was among the first to document the presence of anamorphic endophytes in perennial grasses from South America. The anamorphic endophytes in some accessions of *H. bogdani* Wil. and *H. brevisubulatum* ssp. *violaceum* (Trin.) Link were identified as *Acremonium* species as were the endophytes in some of the

Table 5. Summary of national grass collections surveyed for clavicipitaceous endophytes in the United States.

Generic	Available[a]		Infected		Infection	Source
Collection	Sp.	Acc.	Sp.	Acc.	Range(%)[b]	Refr.
Festuca	49(36)	1,666(190)	16	41	NA	Springer(24)
Hordeum	24(17)	27,038(96)	3	17	47-100	Wilson(30)
Lolium	8(8)	961(85)	5	28	1-99	Wilson(29)
Poa	49(49)	760(51)	5	6	NA	K.Clay(pers. comm.)

[a]Values in parentheses indicate the number of species and accessions examined in the survey of each generic collection.
[b]NA, data not available.

Lolium species. A higher incidence of seed infection was found in *Hordeum* accessions from wet, uncultivated habitats than in dry, cultivated sites. Endophyte viability in *Hordeum* seeds also tended to decrease over time in storage. Similarly, seed viability appeared to decrease more rapidly over time in endophyte-infected accessions than in endophyte-free accessions. These results could suggest that endophytes may consume food reserves or accumulate toxic metabolites in seeds over long-term storage that may reduce seed viability and perhaps endophyte viability.

Morphological studies of *Acremonium* endophytes from wild *Hordeum* species showed that these fungi have very similar morphologies to *Acremonium* endophytes in forage and turf grasses. For example, an endophyte from *H. bogdani* produced an abundance of aerial synnemata in culture in the absence of host tissue. Similar synnemata were formed in culture by *A. coenophialum* Morgan-Jones & Gams on autoclaved tall fescue seedling tissue (28). The *H. bogdani* endophyte tended to gradually lose the ability to form organized synnemata as the endophyte was subcultured. Conidiophores were produced abundantly, mostly perpendicular to synnematal hyphal strands (Fig. 2). Other strains produced conidiophores singly in smooth, relatively appressed colonies. Most *Acremonium* endophytes from wild *Hordeum* species produced abundant conidia in culture. Conidia often gave rise to seed colonies around the mother colony in older cultures. Conidia were elliptical to cresent-shaped forming singly on determinant conidiophores. Some strains puckered the agar similarly to other anamorphic endophytes such as *Acremonium lolii* Latch,Christensen & Samuels.

Further testing of plant inventory accessions has differentiated genetic resistance from endophyte-associated resistance to several major cereal insect pests. Clement *et al.* (5, 6) have shown that the presence of anamorphic endophytes in grasses can significantly impact the biology of the Russian wheat aphid, *Diuraphis noxia* (Mordvilko). This aphid is an exotic pest introduced into the U.S. (Texas) from Russia probably via Mexico in 1986 (25). The aphid has subsequently spread northward throughout the Plain states and westward to the Pacific Northwest where it significantly reduces wheat and barley production. The bases of resistance associated with each endophyte-host and cereal aphid species are presented in Table 6. Most endophyte-associated resistance is due to antibiosis. However, resistance in fescue accessions to the oat birdcherry aphid (*Rhopalosiphum padi* L.) is probably due to antixenosis. The response and poor fecundity of the Russian wheat aphid on several wild grasses make it possible to use the aphid as an assay tool to detect endophytes in infected plants (29).

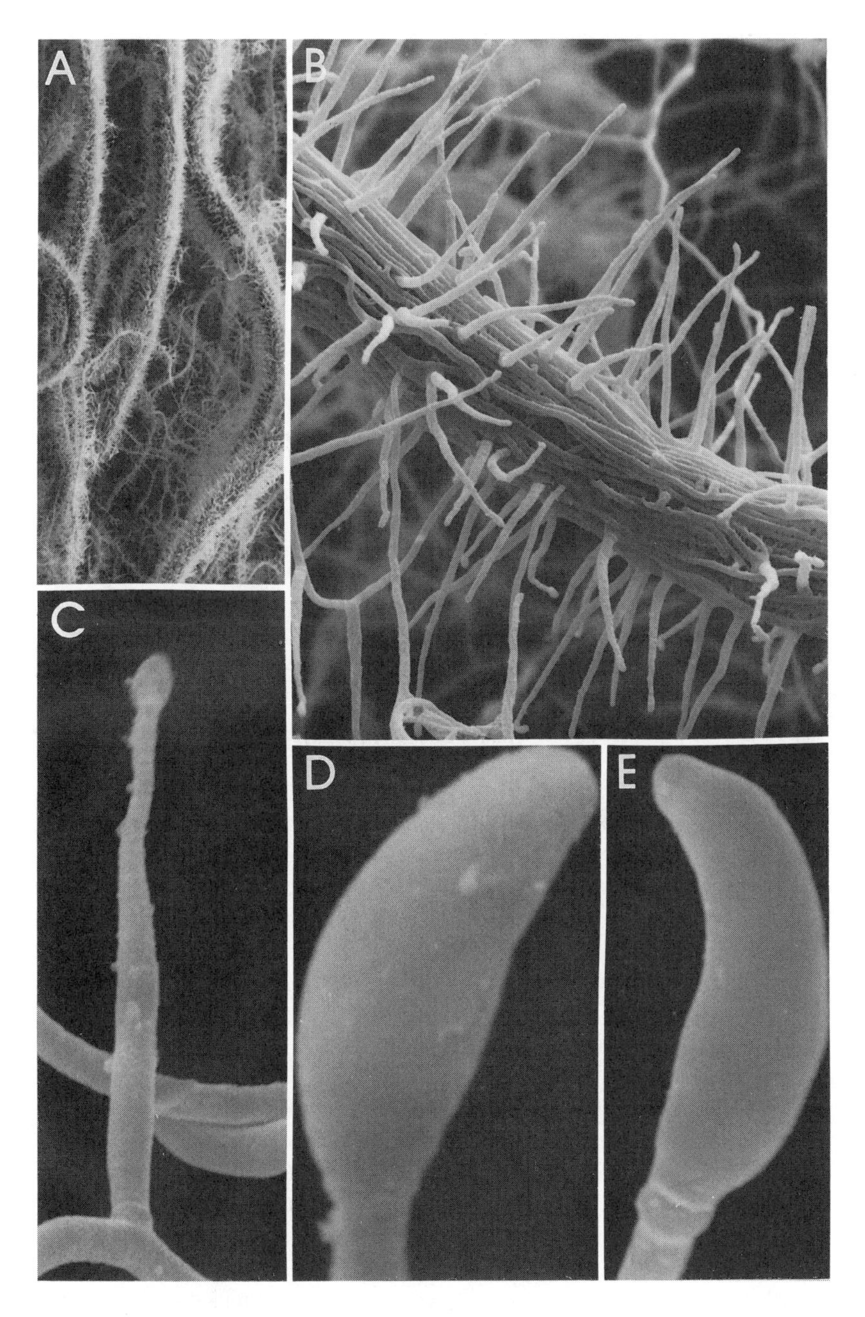

Fig. 2. Scanning electron micrographs of *Acremonium* endophyte from *H. bogdani*. (a) Synnemata, × 70 (b) Synnema, × 1170 (c) Conidiophore, × 4710 (d) Conidium, × 18480 (e) Conidium, × 11850.

Livestock Toxicoses Caused by Endophytes

Much of the initial interest in endophyte research arose with discoveries of associations between fungal endophytes of grasses and animal toxicoses. Bacon *et al.* (1) in 1977 reported the close relationship between an endophyte of tall fescue (*Festuca arundinacea* Schreb.) and incidence of fescue toxicosis (summer syndrome) of cattle in the eastern United States. Fletcher and Harvey (9) reported a similar association in 1983 between an endophyte of perennial ryegrass and ryegrass staggers of sheep in New Zealand. Further studies showed that specific alkaloids such as the tremorgenic lolitrem neurotoxins and ergot alkaloids were responsible for

Table 6. Bases of resistance in endophyte-infected national grass germplasm to cereal aphids.

Cereal Aphid Species	Host Plant Species	Endophyte Species	Resistance Mechanism	Reference
Diuraphis noxia	*Festuca arundinacea*	*A. coenophialum*	Antibiosis	Clement(5)
	Lolium perenne	*A. lolii*	Antixenosis	Clement(6), Wilson (29)
			Antibiosis	Clement(5)
	Hordeum brevisubu-latum	*Acremonium* sp.	Antibiosis	UD
Rhopalosiphum padi	*Festuca arundinacea*	*A. coenophialum*	Antixenosis	UD

UD = unpublished data.

these livestock maladies (10, 11, 19, 35). The alkaloids were found to be systemically distributed in grass seeds and tillers consumed by livestock. Endophytes also were found to produce other types of alkaloids including peramine, lolines (pyrrolizidines), norharmane, halostachine, and perloline (34), some of which confer resistance to certain insect pests and some have relatively low apparent toxicities to mammalian herbivores.

Research on livestock toxicoses caused by clavicipitaceous endophytes in the U.S. has mostly concentrated on teleomorphic endophytes of the Balansieae (Ascomycetes) and fescue toxicoses caused by the anamorphic endophyte *A. coenophialum* in the eastern states. The tall fescue endophyte causes hair loss, reduced milk production, reduced weight gain, and gangrenous dry rot of hooves (fescue foot) in cattle (22, 23). The ryegrass endophyte causes staggers in both sheep and cattle in the United States. Ryegrass staggers and fescue toxicosis likewise are frequently reported in the western states. These diseases appear more regularly in grass-seed producing areas, but livestock toxicoses have been reported throughout the west. A summary of some endophyte-associated toxicoses diagnosed by the author since 1989 are provided in Table 7. Fescue foot is probably the most observed western disease of cattle that consume endophyte-infected tall fescue.

Tests of endophyte-infected accessions in germplasm collections have permitted identification of potentially hazardous materials before distribution. Field surveys of endophyte-infected wild grasses should reveal new livestock toxicoses. A new hyperthermia syndrome of cattle apparently caused by an *Acremonium* endophyte of perennial ryegrass was recently reported in coastal pastures of Pacific Co. in Washington State (12, 33). Temperatures were elevated in cattle of all ages throughout all seasons in affected pastures, but were higher in July than in May, October, and January. Calves had higher temperatures than cows grazing on the same pasture. Cattle removed from affected pastures and fed alfalfa became normothermic within several days, suggesting that the hyperthermia was caused by a pyrogenic factor in the feed. The endophyte appeared morphologically distinct from *A. lolii* and *A. typhinum* Morgan-Jones & Gams (=*Sphacelia typhina* (Pers.) Sacc.), and grew faster in culture than most anamorphic endophytes. Other diseases caused by clavicipitaceous endophytes will no

Table 7. Examples of livestock toxicoses caused by consumption of *Acremonium* endophyte-infected forage grasses in the Pacific Northwest.

Toxicosis	Stock	Location	Endophyte	Toxin	References[a]
Ryegrass staggers	cattle	Payette Co., ID	*A. lolii*	Lolitrem neurotoxin	UD
	sheep	Benton Co., OR	*A. lolii*	Lolitrem neurotoxin	UD
Fescue Foot	cattle	Benton Co., WA	*A. coeno-phialum*	Ergot Alkaloids	UD
Ryegrass fever	cattle	Pacific Co. WA	*Acremonium* spp.	unknown	Wilson (33)

[a]UD = unpublished data.

doubt be discovered from field surveys for endophytes in forage grasses.

Conclusions

There are vast graminaceous germplasm resources in the NPGS available for surveys and testing of endophyte-based resistance to pests and useful agronomic traits. Portions of some generic grass collections have been examined for endophyte infections, but the majority of NPGS grass collections remain to be tested. Many very large collections will require extensive investigations to be adequately evaluated. The rapidly expanding germplasm resources in the already extensive NPGS inventory increase the need to test grass germplasm before introduction into NPGS inventory. This is particularly true because these resources are maintained by limited personnel with finite resources. Increasing reports of endophyte-associated livestock toxicoses in the United States and abroad are other examples where pretesting is needed before introduction and distribution of potentially toxic germplasm occurs.

Previous surveys of endophyte incidence in generic U.S. grass germplasm collections indicate that endophyte-infected accessions usually comprise a relatively small portion of any one collection (29). Endophytes are present in many generic grass collections, but they tend to occur at relatively low incidences and low rates of infection in most NPGS collections compared with commercial collections. The rarity of infected accessions in national generic grass collections and the value of this material suggest that infected accessions should be given some special care in germplasm maintenance programs to maintain endophyte viability and diversity.

The resistance that endophytes provide against graminaceous pests is independent of genetic resistance. Consequently, endophyte-infected accessions are potentially as valuable as accessions with genetic resistance. The diversity of endophyte strains among infected accessions provides a wide variety of types and amounts of alkaloids and other secondary metabolites that may be utilized in controlling specific pests or in producing desired levels of agronomic traits. The combined use of endophyte-based resistance and genetic resistance permits additional alternatives in integrated approaches to pest management. Presently, anamorphic endophytes are used primarily to suppress pests of turfgrasses and reduce populations of soilborne pathogens in certain leguminous crops via crop rotations with endophyte-infected, graminaceous trap crops. The direct use of endophytes as sources of resistance in turfgrasses might be expanded to grass crops by the creation of new endophyte-grass combinations using artificial inoculation methods developed by Latch *et al.* (14). This may be possible as indicated previously because endophytes produce some alkaloids that are toxic to pests but innocuous to mammals, perhaps including man. Endophyte-infected forage grass cultivars already are being developed in

New Zealand to provide resistance to insect pests, but with low toxicity to livestock (Latch, personal communication). Since grass crops produce the vast majority of the food consumed in the world, the utility of clavicipitaceous anamorphic endophytes as new potential sources of resistance to pests of these crops should be thoroughly explored. It is apparent that the economic importance of fungal endophytes to agriculture should encourage plant scientists to greater utilize and test these valuable national germplasm resources.

Literature Cited

1. Bacon, C. W., Porter, J. K., Robbins, J. D., and Luttrell, E. S. 1977. *Epichloë typhina* from toxic tall fescue grasses. Appl. Environ. Microbiol. 34:576-581.
2. Clay, K. 1988. Clavicipitaceous endophytes of grasses: coevolution and the change from parasitism to mutualism. Pages 79-105 in: Coevolution of Fungi with Plants and Animals. D. Hawksworth & K. Pirozynski, eds. Academic Press, London.
3. Clay, K. 1989. Clavicipitaceous endophytes of grasses: their potential as biocontrol agents. Mycol. Res. 92:1-12.
4. Clay, K. 1990. Fungal endophytes of grasses. Annu. Rev. Ecol. Syst. 21:275-297.
5. Clement, S. L., Pike, K. S., Kaiser, W. J., and Wilson, A. D. 1991. Resistance of endophyte-infected plants of tall fescue and perennial ryegrass to the Russian wheat aphid (Homoptera: Aphididae). J. Kans. Entomol. Soc. 63:646-648.
6. Clement, S. L., Lester, D. G., Wilson, A. D., and Pike, K. S. 1992. Behavior and performance of *Diuraphis noxia* (Homoptera: Aphididae) on fungal endophyte-infected and uninfected perennial ryegrass. J. Econ. Entomol. 85:583-588.
7. De Battista, J. P., Bacon, C. W., Severson, R., Plattner, R. D., and Bouton, J. H. 1990. Indole acetic acid production by the fungal endophyte of tall fescue. Agron. J. 82:878-880.
8. Diehl, W. W. 1950. *Balansia* and the Balansiae in America. Agric. Monogr. 4., USDA, Washington, DC. 82 pp.
9. Fletcher, L. R., and Harvey, I. C. 1981. An association of a *Lolium* endophyte with ryegrass staggers. N. Z. Vet. J. 29:185-186.
10. Gallagher, R. T., Hawkes, A. D., Steyn, P. S., and Vleggaar, R. 1984. Tremorgenic neurotoxins from perennial ryegrass causing ryegrass staggers disorder of livestock: structure and elucidation of lolitrem B. J. Chem. Soc. Chem. Commun. 9:614-616.
11. Gallagher, R. T., White, E. P., and Mortimer, P. H. 1981. Ryegrass staggers: isolation of potential neurotoxins lolitrem A and lolitrem B from staggers-producing pastures. N. Z. Vet. J. 24:189-190.

12. Gay, C. C., Fransen, S. C., Wilson, A. D., Boyes, P. E., Waller, S. L., and Kinsell, M. L. 1991. Enzootic hyperthermia in cattle in a coastal region of Pacific County Washington State. Proceedings of the 12th Annual Food Animal Disease Research Conference. University of Wyoming,.Laramie, WY. (Abstract).
13. Latch, G. C. M., Christensen, M. J., Samuels, G. J. 1984. Five endophytes of *Lolium* and *Festuca* in New Zealand. Mycotaxon 20:535-550.
14. Latch, G. C. M., and Christensen, M. J. 1985. Artificial infection of grasses with endophytes. Ann. Appl. Biol. 107:17-24.
15. Latch, G. C. M., Potter, L. R., and Tyler, B. F. 1987. Incidence of endophytes in seeds from collections of *Lolium* and *Festuca* species. Ann. Appl. Biol. 111:59-64.
16. Lindau, G. 1904. Über das Vorkommen des Pilzes des Taumellolchs in altägyptischen Sämen. Sitzungsber. K. Preuss. Akad. Wissen, pp. 1031-1036.
17. Morgan-Jones, G., and Gams, W. 1982. Notes on hyphomycetes. XVI. An endophyte of *Festuca arundinacea* and the anamorph of *Epichloë typhina*, new taxa in one of two new sections of *Acremonium*. Mycotaxon 15:311-318.
18. Philipson, M. N. 1989. A symptomless endophyte of ryegrass (*Lolium perenne*) that spores on its host - a light microscope study. N. Z. J. Bot. 27:513-519.
19. Prestidge, R. A., and Gallagher, R. T. 1985. Lolitrem B - a stem weevil toxin isolated from *Acremonium*-infected ryegrass. Proc. N. Z. Weed Pest Control Conf. 38:38-40.
20. Saha, D. C., Johnson-Cicalese, J. M., Halisky, P. M., van Heemstra, M. I., and Funk, C. R. 1987. Occurrence and significance of endophytic fungi in the fine fescues. Plant Dis. 71:1021-1024.
21. Shands, H. L. 1990. Plant genetic resources conservation: The role of the gene bank in delivering useful genetic materials to the research scientist. J. Heredity 81:7-10.
22. Siegel, M. R., Latch, G. C. M., and Johnson, M. C. 1985. *Acremonium* fungal endophytes of tall fescue and perennial ryegrass: significance and control. Plant Dis. 69:179-183.
23. Siegel, M. R., Latch, G. C. M., and Johnson, M. C. 1987. Fungal endophytes of grasses. Ann. Rev. Phytopathol. 25:293-315.
24. Springer, T. L., and Kindler, S. D. 1990. Endophyte-enhanced resistance to the Russian wheat aphid and the incidence of endophytes in fescue species. Pages 194-195 in: Proc. 2nd Intern. Sympos. on *Acremonium*/Grass Interactions. S. S. Quisenberry and R. E. Joost, eds. Louisiana Agricultural Experiment Station, Baton Rouge.
25. Stoetzel, M. B. 1987. Information on and identification of *Diuraphis noxia* (Homoptera: Aphididae) and other aphid species colonizing

leaves of wheat and barley in the United States. J. Econ. Entomol. 80:696-704.

26. Vogl, A. E. 1898. Mehl und die anderen Mehlprodukte der Cerealien und Leguminosen. Nahrungsm. Unters. Hyg. Warenk. 12:25-29.
27. White, J. F., Jr. 1987. Widespread distribution of endophytes in the Poaceae. Plant Dis. 71:340-342.
28. White, J. F., Jr., and Cole, G. T. 1985. Endophyte-host associations in forage grasses. II. Taxonomic observations on the endophyte of *Festuca arundinacea*. Mycologia 77:483-486.
29. Wilson, A. D., Clement, S. L., and Kaiser, W. J. 1991. Survey and detection of endophytic fungi in *Lolium* germplasm by direct staining and aphid assays. Plant Dis. 75:169-173.
30. Wilson, A. D., Clement, S. L., Kaiser, W. J., and Lester, D. G. 1991. First report of clavicipitaceous anamorphic endophytes in *Hordeum* species. Plant Dis. 75:215.
31. Wilson, A. D., Clement, S. L., and Kaiser, W. J. 1991. Endophytic fungi in a wild *Hordeum* germplasm collection. FAO/IBPGR Pl. Gen. Res. Newsl. 87:1-4.
32. Wilson, A. D., Kaiser, W. J., and Clement, S. L. 1991. Clavicipitaceous endophytes in wild *Hordeum* germplasm. Phytopathology 81:1151. (Abstract).
33. Wilson, A. D., Gay, C. C., and Fransen, S. C. 1992. An *Acremonium* endophyte of *Lolium perenne* associated with hyperthermia of cattle in Pacific County, Washington. Plant Dis. 76:212.
34. Yates, S. G., and Powell, R. G. 1988. Analysis of ergopeptine alkaloids in endophyte-infected tall fescue. J. Agric. Food Chem. 36:337-340.
35. Yates, S. G., Plattner, R. D., and Garner, G. B. 1985. Detection of ergopeptine alkaloids in endophyte-infected toxic Ky-31 tall fescue by mass spectrometry/mass spectrometry. J. Agric. Food Chem. 33:719-722.

CHAPTER 10

NATURAL VARIATION AND EFFECTS OF ANTHROPOGENIC ENVIRONMENTAL CHANGES ON ENDOPHYTIC FUNGI IN TREES

Marjo L. Helander, Seppo Neuvonen and Hanna Ranta

Department of Biology and Kevo Subarctic Research Institute
University of Turku
FIN-20500 Turku, Finland

Environmental changes can influence plants by altering interactions between microbial symbionts (e.g. endophytic fungi) and plant pathogens and herbivores. Because endophytic fungal infections seem to be widespread, these interactions deserve more attention than they have received to date.

The ecological roles of endophytic fungi are varied and may change during their life cycle. They may be dormant saprobes, latent pathogens or mutualistic symbionts e.g. antagonizing herbivores and plant pathogens. There is some evidence that endophytes may be part of the defense system of the tree against pathogens and pests (5, 15). They may act directly as antagonists or indirectly by inducing host metabolic responses against plant pathogens and herbivores.

Fungi known to be potential pathogens can live for a certain period of their life cycle as neutral endophytes and cause symptoms only after specific ecological and physiological conditions (24, 26). Latency can last from several days to many years depending on fungal virulence, physiological condition of the host, climate and environmental stresses e.g. air pollution.

Patterns in the Occurrence of Endophytic Fungi

Extensive studies of endophytes of woody perennials during the last two decades have produced some information on the general patterns of endophytic occurrence in trees, especially in conifers (5, 15, 19). Endophytes seem to be distributed in patches even within single branches (7). This patchiness may be very important for the plant interactions with herbivores and pathogens. Endophytes might differentially modify leaf quality to consumers and so transform trees into mosaics of favorable and unfavorable substrates.

Infection rates within a tree tend to increase with increasing foliage age and decreasing distance from the trunk (20, 27). Continued exposure to inoculum accounts for increasing rates of infection with age of the foliage. A more humid microclimate accounts for increased infection rate within the crown compared to the edge of the crown. Incidence of fungal endophytes decreases with increasing height of the canopy (11, 27).

The frequency of colonization by endophytic fungi decreases with increasing altitude in Norway spruce (*Picea abies* (L.) Karst.) needles (23) and by low temperatures in annual shoots of Scots pine (*Pinus sylvestris* (L.)) and Norway spruce (3). Samples from homogenous stands with a closed canopy showed higher overall infection rates than those from mixed stands with an open canopy (27). Endophytes also occur more frequently in moist than in dry sites (4). For a host-specific endophyte the inoculum potential increases as the density of the canopy and the homogeneity of the host stand increases.

From studies with grass endophytes it seems evident that the degree of mutualism between host and endophyte is dependent on the plant environment. When plants grow under stressful conditions such as low nutrient levels (6) or low light intensity (14), uninfected plants tend to do as well or better than infected plants. An important question for further studies is in which kind of environmental situations bearing endophytes is beneficial for trees, i.e. what determines when the relationship is mutualistic.

Pollution and Microfungi of Above Ground Parts of Plants

Air pollution affects trees directly by damaging needles and leaves and causing a decrease in the assimilation capacity of the canopy. Indirect effects occur via ground, when e.g. acid rain changes the nutrition content of the soil and causes accumulation of hazardous ions. Microfungi living on or in aerial plant parts may also be affected via both of these pathways, directly and indirectly. If air pollution changes the quality or species composition of epi- and endophytic fungi, that may have various consequences for other parts of the ecosystem e.g. on host plant, plant pathogens and herbivorous insects.

Experimental field studies, gradient analyses and laboratory experiments have shown that atmospheric pollution affects phyllosphere microbial communities, which are directly exposed to atmospheric pollutants (13). These effects are short term influences (8) and the recovery may be fast. By contrast, endophytic fungi live most of their life cycle in an environment which is protected against sudden weather changes and other environmental factors e.g. air pollution. Dark pigmentation protects some fungi from the harmful effects of UV-light. However, endophytic fungi generally lack the thick walled, melanised structures present in the vegetative hyphae of epiphytic fungi (18).

Erosion of waxes from leaf surfaces by air pollutants modifies the microhabitat of germinating fungal spores (1) and may cause changes in hyphal penetration to inner plant tissues. Changes in leaf surface ultrastructure may also have an effect on persistence of canopy wetness, which in turn affects fungal spore germination and growth.

Damaged stands are often irregularly spaced, leaving frequent gaps in the vegetation. It has been shown (21) that densely wooded sites with evergreen shrubs had higher overall levels of infection by endophytes than less dense sites. This may be due to more favourable moisture conditions for endophytes in dense stands. Changes in endophyte infection rates may thus be a result of other environmental changes rather than direct effects of air pollution on endophytes.

Endophyte frequency on Norway spruce (23) and white fir (*Abies alba* Mill.) (24) stands has been studied, but significant differences between healthy and damaged sites could not be detected. Different climatic conditions may be responsible for the large differences among the populations of endophytic fungi at different study sites. Occurrence of the endophytic fungus *Diaporthe eres* Nitschke on beech (*Fagus sylvatica* L.) leaves was significantly higher in leaves of healthy trees than those of damaged ones (25).

Four *Lophodermium* species in litter of Scots pine needles responded to high levels of air pollution in the surroundings of an industrial town in southwestern Finland (10). The amount of endophytic fungi in living needles at the same sites has not been studied. The hysterothecia production of these ascomycete species was related to the distance from the factory complex producing copper, nickel, sulphuric acid and fertilizers, and to the chemical composition of living needles. The hysterothecia were almost totally absent within a distance of 800 m from the emission source (copper concentrations 55-898 ppm). The pattern was rather similar in all the studied fungal species, but the adverse effect of air pollution was clearest in the most abundant species, *L. pinastri* (Schrad.:Fr.) Chev. The decrease of the *Lophodermium* species is probably due to the toxicity of the industrial emissions (especially heavy metals, whose fungicidal properties are well known), but indirect effects, such as impoverished vegetation and associated changes in the microclimate may have played an additional role.

Ascocarps of *Lophodermium piceae* (Fuckel) Höhn., an endophytic fungus of Norway spruce needles, normally develop after the infected needles have fallen to the ground, and they mature during autumn. Initial spore infection by *L. piceae* seems to take place mainly during the first growing season of the Norway spruce needles. Its high incidence within needles (about 90 %) seemed to be typical for young, healthy stands (2). The occurrence of *L. piceae* was lower in both young and old damaged stands, showing higher rates of needle loss than in young healthy ones. Differences in infection levels between damaged stands, and similar levels of infection in healthy and damaged trees within the same stand indicate that other factors, e.g. environmental or genetic, are more important than damage in determining *L. piceae* incidence (2).

A Case Study: Effects of Simulated Acid Rain on Endophytic Fungi of Pine Needles

In earlier published studies the effects of atmospheric pollution on quantity and species composition of endophytic fungi have mainly been studied by comparing damaged and undamaged stands or individual trees. In these situations it is difficult to exclude naturally occurring variance between the study sites caused by other environmental factors than air pollution.

The patchiness of endophytic distribution causes an additional problem when studying effects of environmental factors on endophytic fungi. One should have sufficient number of study sites, sampled trees and samples within a tree.

The aim of our study was to experimentally investigate whether simulated acid rain (SAR) has effects on amount of endophytic fungi inhabiting Scots pine (*Pinus sylvestris* L.) needles. We also investigated the effect of nearby pine tree density on the occurrence of needle endophytes in our study plots.

Study area and experimental set-up

Our study area is situated in northernmost Finland (69°45'N, 27°E), where background pollution values are low (12) but environmental conditions are otherwise harsh. The experiment consisted of 80 study plots. The study area was divided into four subareas, 20 plots on each. Subareas were further divided into 5 blocks of four plots, and within each block allocated randomly to four treatments. In 1985-1988 all subareas had the following four treatments: dry control, irrigated control and two levels (pH 4 and pH 3) of combined sulphuric and nitric acid treatments. Since ambient rain was not excluded from the plots, the treatments can be considered to simulate a situation where "acid rain" episodes are interspersed among natural rainfall events. In 1989 the experiment was modified as follows: in subarea 2 only sulphuric acid and in subarea 3 only nitric acid was applied to the acid treatment plots; in subareas 1 and 4 the experiment was continued unchanged.

The water for the acid treatments was prepared by adding sulphuric and/or nitric acid (with sulphate (SO_4^-) to nitrate (NO_3^-) ratio of 1.9:1 by weight. The irrigation treatments were conducted three times a week by sprinklers over the tree canopy and a 5 m x 5 m area of ground, 5 mm per occcasion for all irrigation treatments, during the growing season (June to September). More detailed description of the experimental set-up, treatments, vegetation and background deposition levels is given elsewhere (9, 16, 17).

Three year old needles were collected from pines from all of the 80 study plots in the beginning of August 1989 and at the end of July 1990. In the first study year 12 needles and in the second year 20 needles were taken from each study tree. Needles were processed immediately or stored at 4 °C before handling. Needle samples were surface sterilized (1 min in 76 % alcohol, 3 min in 8% natrium hypochlorite, 0.5 min in 76 % alcohol, rinsed twice in distilled sterilized water) and cut longitudinally into two sections. The two segments were placed with the cut half in contact with the malt extract (2%) agar surface and incubated at room temperature. Colonies were counted once a week for two months. Endophytes were grouped into two ecologically different groups, those emerging from the needle base or lamina. Total amount of commonly occurring *Leptostroma* sp. was also counted.

Results

Most (65 %) of the isolated endophytes emerged from the lamina of the needle. Almost half of the total amount of endophytes belonged to *Leptostroma* sp. Overall, 1.5 endophytic fungi emerged per 10 needles at our study area. There was, however, wide variation among individual trees: from 0 to 7.9 endophytes per 10 needles. A significant proportion of this variation was explained by the density of pines surrounding the study trees (Fig. 1).

Since needles from Scots pines having few other pines in their vicinity beared none or only few endophytic fungi, we excluded plots with no pines within 5 m of the study tree when analysing the effects of treatments on the abundance of endophytes. Furthermore, the density of surrounding pines was used as a covariate in these analyses. Subarea 3 was analysed separately from other subareas since the overall density of pines was considerably higher in this subarea and the pH levels of the acid treatments in 1989 and 1990 were higher (due to the exclusion of sulphuric acid) than in other subareas (9).

There were significant differences in the amount of endophytic fungi among different subareas and due to the density of surrounding pines (Tables 1 and 2; Fig. 1). In 1989 the endophyte amounts did not show any relationship to the simulated acid rain treatments (Fig. 2; Tables 1 and 2). In 1990 on subareas having sulphuric acid or both sulphuric and nitric acids in SAR treatments there was a tendency ($p < 0.10$) of differences among treatments (Table 1). On these subareas the amounts of endopytic fungi in pine needles were consistently lowest in the stronger acid rain treatment in 1990 (A3; Fig. 2). The mean abundance of endophytes in A3 was 64 % lower than that in irrigated controls.

Table 1. Analysis of variance for the differences among subareas (1(S+N), 2(S) and 4(S+N)), the effects of simulated acid rain and the density of surrounding (< 5 m) pine trees (as a covariate) on the amounts of endophytic fungi of Scots pine needles in 1989 and 1990 (E-89 and E90, respectively). Plots with no other pines within 5 m of the study tree were excluded from the analysis. p = risk probability, EMS = error mean square.

		Log(E-89)		Log(E-90)	
Source	DF	F	p	F	p
Subarea (SA)	2	9.32	0.001	5.70	0.008
Treatment (TRT)	3	0.61	0.616	2.54	0.076
SA*TRT	6	1.12	0.377	0.27	0.947
Log(trees)	1	11.58	0.002	6.95	0.014
Error	28 EMS	0.2508417		0.1506980	

Table 2. Analysis of variance for the effects of simulated acid rain and the density of surrounding (< 5 m) pine trees on the amounts of endophytic fungi of Scots pine needles in subarea 3 (N). Plots with no other pines within 5 m of the study tree were excluded from the analysis.

		Log(E-89)		Log(E-90)	
Source	DF	F	p	F	p
Treatment	3	0.38	0.766	0.56	0.649
Log(trees)	1	1.68	0.220	6.77	0.023
Error	12 EMS	0.4866615		0.3463801	

Similar comparisons of epiphytic fungi on pine needles showed significant and larger differences due to SAR treatment (9, 22). Endophytic fungi are protected from the effects of environmental changes e.g. air pollution, compared with epiphytic micro-organisms. However, if endophytic communities are affected by a long term exposure to pollutants, the change might be more permanent.

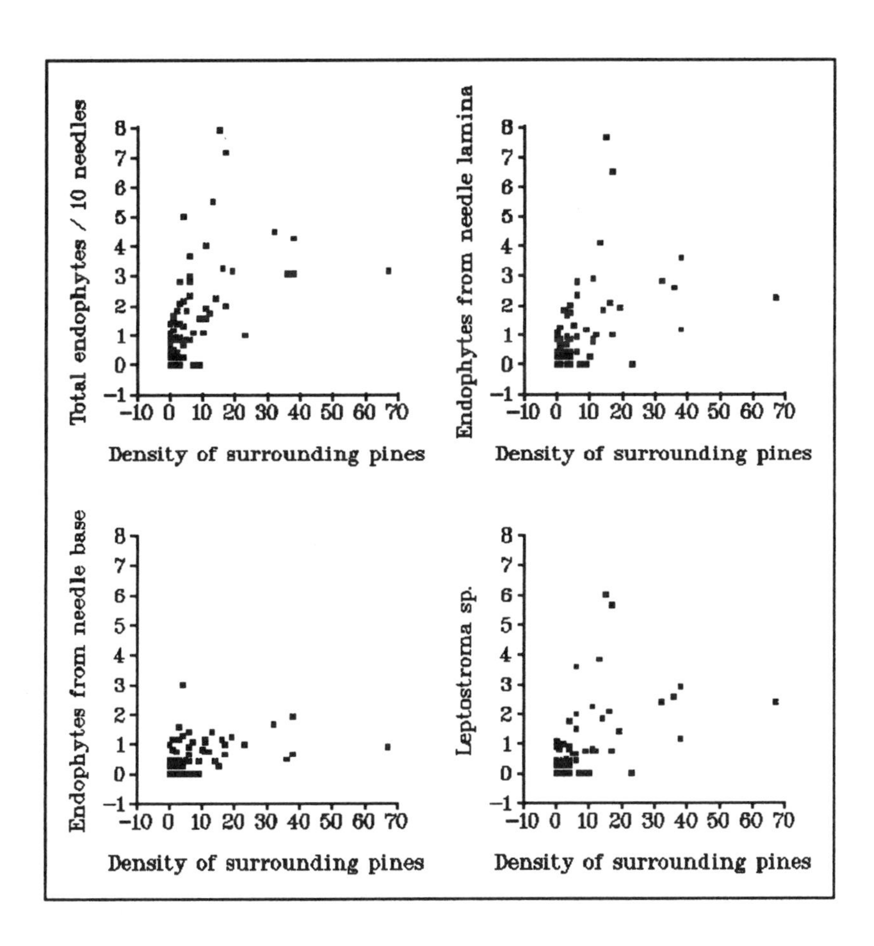

Fig. 1. The abundance of endophytic fungi in Scots pine needles (unweighted tree specific means of data from 1989 and 1990) in relation to the density of surrounding pine trees (number of other pines within a circle with a 5m radius). Separate figures are shown for all endophytes, those emerging from needle lamina and needle base, and for *Leptostroma* sp.

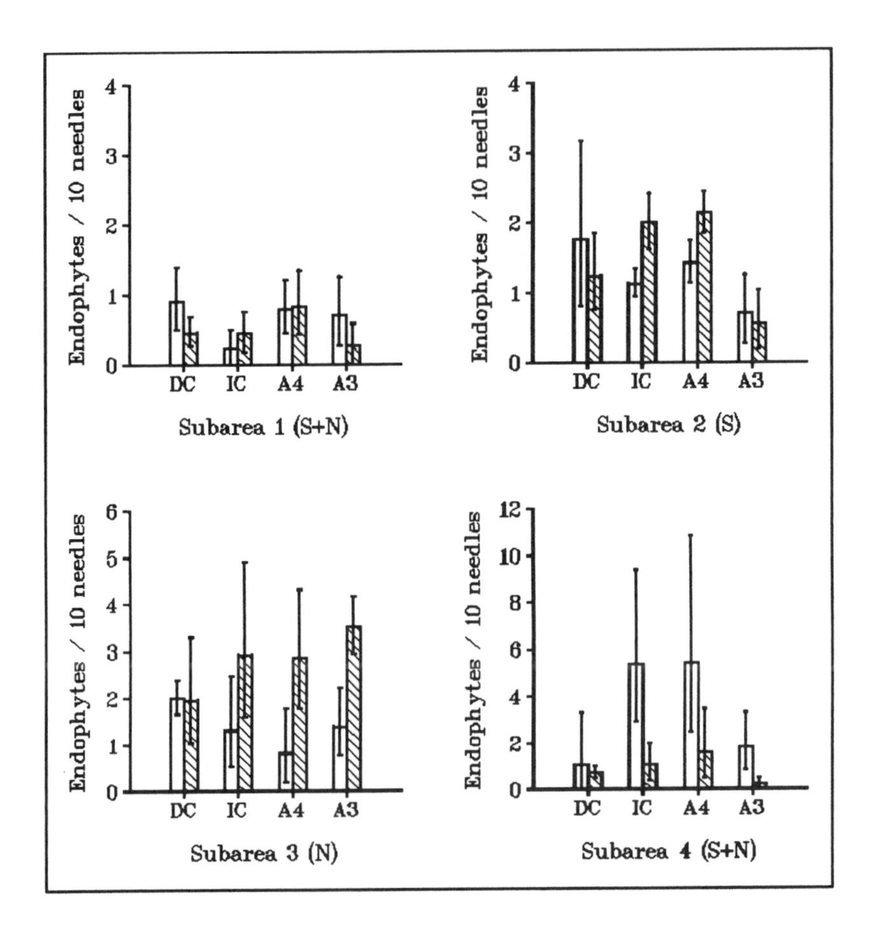

Fig. 2. The abundance of endophytic fungi in Scots pine needles in different subareas and treatments. Treatments: DC = dry control, IC = irrigated control, A3 = simulated acid rain with stronger acid load, A4 = simulated acid rain with smaller acid load; the letters in parentheses show whether the acid rain treatments had sulphuric acid (S), nitric acid (N) or both (S+N) during 1989 and 1990; in all subareas both (S+N) were applied in SAR treatments from 1985 to 1988. Open bars: August 1989; hatched bars: July 1990. Mean ± S.E. are shown, values are back transformed from logarithms and consequently the error bars are asymmetric.

Acknowledgements

We thank Dr. Naresh Magan for helpful comments on the manuscript. This study was supported by the Academy of Finland and Maj & Tor Nessling Foundation.

Literature Cited

1. Allen, E.A., Hoch, H.C., Steadman, J.R. and Stavely, R.J. 1991. Influence of leaf surface features on spore deposition and the epiphytic growth of phytopathogenic fungi. Pages 87-110 in: Microbial Ecology of Leaves. J.H. Andrews and S. Hirano, eds. Springer-Verlag, New York.
2. Barklund, P. 1987. Occurrence and pathogenicity of *Lophodermium piceae* appearing as an endophyte in needles of *Picea abies*. Trans. Br. Mycol. Soc. 89:307-313.
3. Barklund, P. and Unestam, T. 1988. Infection experiments with *Gremmienella abietina* on seedlings of Norway spruce and Scots pine. Eur. J. For. Pathol. 18:409-420.
4. Carroll, G.C. and Carroll, F.E. 1978. Studies on the incidence of coniferous needle endophytes in the Pacific Northwest. Can. J. Bot. 56:3034-3043.
5. Carroll, G.C. 1986. The biology of endophytism in plants with particular reference to woody perennials. Pages 205-222 in: Microbiology of the Phyllosphere. N.J. Fokkema and J. Van den Heuvel, eds. Cambridge University Press, Cambridge.
6. Cheplick, G.P., Clay, K. and Marks, S. 1989. Interactions between infection by endophytic fungi and nutrient limitation in the grasses *Lolium perenne* and *Festuca arundinacea*. New Phytol. 98:165-175.
7. Espinosa-Garcia, F.J. and Langheim, J.H. 1990. The endophytic fungal community in leaves of a coastal redwood population - diversity and spatial patterns. New Phytol. 116:89-97.
8. Helander, M.L. and Rantio-Lehtimäki, A. 1990. Effects of watering and simulated acid rain on quantity of phyllosphere fungi of birch leaves. Microb. Ecol. 19:119-125.
9. Helander, M.L., Ranta, H. and Neuvonen, S. 1993. Responses of phyllosphere microfungi to simulated nitric and sulphuric acid deposition. Mycol. Res. 97:533-537.
10. Heliövaara, K., Väisänen, R. and Uotila, A. 1989. Hysterothecia production of *Lophodermium* species (Ascomycetes) in relation to industrial air pollution. Karstenia 29:29-36.
11. Johnson, J.A. and Whitney, N.J. 1989. An investigation of needle endophyte colonization patterns with respect to height and compass direction in a single crown of balsam fir (*Abies balsamea*). Can. J. Bot. 67:723-725.

12. Laurila, T., Tuovinen, J.-P. and Lättilä, H. 1991. Lapin ilmansaasteet ("Air pollution in Lapland", in Finnish). Finnish Meteorological Insitute. Helsinki.
13. Magan, N. and McLeod, A.R. 1991. Effects of atmospheric pollutants on phyllosphere microbial communities. Pages 379-400 in: Microbial Ecology of Leaves. J. H. Andrews and S. Hirano, eds. Springer-Verlag, New York.
14. Marks, S. and Clay, K. 1990. Effects of CO_2 enrichment, nutrient addition, and fungal endophyte-infection on the growth of two grasses. Oecologia 84:207-214.
15. Miller, J. D. 1986. Toxic metabolites of epiphytic and endophytic fungi of conifer needles. Pages 223-231 in: Microbiology of the Phyllosphere. N. J. Fokkema and J. Van den Heuvel, eds. Cambridge University Press, Cambridge.
16. Neuvonen, S. and Suomela, J. 1990. The effect of simulated acid rain on pine needle and birch leaf litter decomposition. J. Appl. Ecol. 27:857-872.
17. Neuvonen, S. Suomela, J. Haukioja, E. Lindgren, M. and Ruohomäki, K. 1990. Ecological effects of simulated acid rain in a subarctic area with low ambient sulphur deposition. Pages 477-493 in: Acidification in Finland. P. Kauppi, P. Anttila, and K. Kenttämies, eds. Springer-Verlag, Berlin.
18. Parbery, D.G. and Emmet, R.W. 1977. Hypothesis regarding appressoria, spores, survival and phylogeny in parasitic fungi. Revue de Mycologie 41: 429-447.
19. Petrini, O. 1991. Fungal endophytes of tree leaves. Pages 179-197 in: Microbial Ecology of Leaves. J. H. Andrews and S. Hirano, eds. Springer-Verlag, New York.
20. Petrini, O. and Carroll, G. 1981. Endophytic fungi in foliage of some Cupressaceae in Oregon. Can. J. Bot. 59:629-636.
21. Petrini, O., Stone, J. and Carroll, F. E. 1982. Endophytic fungi in evergreen shrubs in western Oregon. A preliminary study. Can. J. Bot. 60:769-789.
22. Ranta, H. 1990. Effect of simulated acid rain on quantity of epiphytic microfungi on Scots pine (*Pinus sylvestris* L.) needles. Environ. Pollut. 67:349-359.
23. Sieber, T. N. 1988. Endephytische Pilze in Nadeln von gesunden und geschädigten Fichten (*Picea abies* (L.) Karsten). Eur. J. For. Pathol. 18:321-342.
24. Sieber, T. N. 1989. Endophytic fungi in twigs of healthy and diseased Norway spruce and white fir. Mycol. Res. 92:322-326.
25. Sieber, T. N. and Hugentobler, C. 1987. Endophytische Pilze in Blättern und Ästen gesunder und geschädigter Buchen (*Fagus sylvatica* L.). Eur. J. For. Pathol. 17:411-425.

26. Sieber, T. N., Sieber- Canavesi F. and Dorworth, C. E. 1990. Endophytic fungi on red alder (*Alnus rubra*) leaves and twigs in British Columbia. Can. J. Bot. 69:407-411.
27. Suske, J. and Acker G. 1990. Host-endophyte interaction between *Lophodermium piceae* and *Picea abies*: cultural, ultrastructural and immunocytochemical studies. Sydowia 42:211-217.

CHAPTER 11

MANIPULATION OF ENDOPHYTIC FUNGI TO PROMOTE THEIR UTILITY AS VEGETATION BIOCONTROL AGENTS

Charles E. Dorworth and Brenda E. Callan

Forestry Canada,
The Pacific Forestry Centre
Victoria, British Columbia, Canada,V8Z 1M5

First-order (I°) biocontrol agents, applied either as mycoherbicides for single event weed control or as classical bioagents which perpetuate themselves for continuing weed control are the most common biocontrol agents. Some endophytic fungi also show promise as second-order (II°) biological control agents of forest weeds. The latter are usually opportunistic fungi which are weak pathogens. Special technology as "formulation" and esoteric application methods must be developed in order that these fungi might prove consistently useful. Species of *Melanconis* are prominent among those which have been tested at the Pacific Forestry Centre as candidate biological control agents of *Alnus rubra*.

Rationale

Competing vegetation or weeds have killed or constrained the growth of newly planted forest trees to an extent that hundreds of thousands of hectares of former forest land are now occupied by non-commercial species. Revegetation of such areas with commercially valuable species is often difficult and costly. The chemical herbicide Vision® (Registered Trademark, Monsanto Canada) is responsible for successful weed control on more than 90% of the forest land treated in British Columbia (6). "Controlled burning" is another extremely valuable component of the forest managers' vegetation management strategy. Both of these options have become topics of criticism by environmentalist groups. The use of chemical herbicides in particular and fire as well is being increasingly limited as a consequence of public reaction (2).

Biological control of weeds in agriculture, through the use of pathogenic fungi, has been attempted on a number of occasions during the past several decades, with several notable successes (10,13,19). A major requirement for the same type of research in forestry has been noted (7,18). To date, very few institutes have committed significant resources to such work. Further, so long as chemicals are available, there is little incentive to fund research which will generate products with relatively small industrial profit margins. This is particularly true with respect to introduced microorganisms which are intended to become self-perpetuating. Considering the paucity of currently available forest vegetation management tools and with the best of those seriously endangered in the long run by public pressures, forest managers are showing increased interest in the use of biological controls.

The Pacific Forestry Centre (PFC) in Victoria, British Columbia has emphasized research in biological control with particular emphasis on the development of indigenous fungi as biocontrol agents of weeds. The majority of these fungi are weak or opportunistic pathogens which are markedly dependent upon optimum environmental conditions to generate epiphytotics in nature. Their employment as industrially acceptable biocontrol agents therefore requires the development of specialized technology with respect to formulation and application. The financial support granted such research worldwide has been, unfortunately, a minor fraction of that invested in the development of chemical herbicides.

Biological control in general can be arbitrarily divided into two major descriptive groups:

1. FIRST-ORDER (I°) BIOCONTROL: This strategy is defined as: Direct application of living agents which reduce the individuals of target pest populations either in number or in vigour, or both.

2. SECOND-ORDER (II°) BIOCONTROL: This strategy is defined as: Manipulation of environmental conditions, the targeted hosts, the indigenous microflora or all of these in order to induce the natural pathogenicity or stimulate the virulence of the native microflora, thereby yielding biological control.

Historically, II° Biocontrol has been applied in agriculture for control of pathogens under the umbrella categories of crop rotation, mulching and incorporation of organic matter into soils, area flooding and others. Respectively, these techniques attenuate pathogen populations either by eliminating nutritional bases, negatively adjusting environmental conditions or by encouraging the development of microflora which are antagonistic to the pathogens to be controlled. Similar possibilities exist with respect to vegetation management. As described under the general category of II° Biocontrol, the equilibrium between the host plant and the potential pathogen is tipped in favour of the pathogen, either by strengthening the latter or weakening the former.

It is our intent at PFC to use only environmentally benign biocontrol agents. The use of weak pathogens, which have co-evolved with their hosts for the most part, requires that methods be developed to promote or accentuate their virulence temporarily in local use situations.

Two benefits are evident in the use of indigenous fungi:

1. Operator Control - Where virulence is promoted or host vigour attenuated through application of stress, as with II° biocontrol, the operator can limit the reaction by controlling the application of a stress factor, either quantitatively or qualitatively, and

2. The biosphere as a whole serves as a buffer or a natural sink wherein the pathogen is soon "reabsorbed" by the system after it increases to levels in excess of those which normally exist when the pathogen and host are in balance.

In the case of inoculations with highly virulent or, especially, exotic microorganisms, it is essential that the operator be able to limit the expansion of the resulting epiphytotic. Some if not most exotic organisms, when accidentally or intentionally introduced, are unable to find a suitable ecologic niche in the new environment and, consequently, they perish. At the other extreme, certain pathogens colonize newly encountered niches to the detriment of native organisms and are not buffered by the biosphere to yield an industrially acceptable population equilibrium. Total lack of natural buffering by the biosphere resulted in uncontrolled spread of the introduced pathogens which cause white pine blister rust, chestnut blight and oak wilt.

The true indigenous, mutualistic endophytes may be regarded as fungi fully buffered within the biosphere and a number of these are being considered at present at PFC as II° candidate biocontrol agents. At the same time, other fungi, which are true necrotrophs, are being tested at PFC in the course of research operations to define and formulate I° candidate biocontrol fungi. The latter work does not involve endophytes and will not be considered further at this time.

An Applied Definition of “Endophyte”

Several authors have proposed definitions of "endophyte" which pointedly exclude plant pathogens from consideration (1,3). Broader definitions of the endophyte concept permit consideration of both pathogenic and mutualistic fungi under the general category "endophyte" (11,16). Sieber *et al.* (17) proposed that the length of the latent endophyte phase is inversely related to the virulence of a pathogen. It may also be postulated that the length of the latent endophytic stage is directly related to the extent of evolutionary advance, or regression from the pathogenic to the mutualistic state.

In all probability there is a significant group of endophytic fungi in healthy plants which become pathogenic when their host plants are weakened. When this occurs, the host-fungus interaction manifests itself as a disease syndrome. Such fungi, often referred to as opportunists, may pass from latent or mutualistic mode to necrotrophic mode when the plant is stressed or "predisposed" (15). Differentiation between opportunistic pathogens and secondary saprophytes *in situ* is difficult and is a topic of debate which is not always open to resolution with extant technology. Chapela and Petrini, in the course of a presentation to the European Weed Research Society (Univ. Lancaster, 1991 April 10-11, personal communication), observed "...all these groups of fungi are not as separate as some might like them to be, but on the contrary, they form a continuum of interactions between the plant and fungus that range from the casual and superficial to the intimate and deeply involved." Our current studies, designed within an applied frame of reference to generate vegetation biocontrol agents from both endophytes and necrotrophs, are bound to contribute fundamental information with respect to the status of both sets of fungi within the pathogen-saprophyte continuum.

Specifically, the directed application of predisposing agents onto the host at increasing concentrations (from sub-lethal through lethal) will permit us to determine which endophytes begin to proliferate and act as pathogens, and to direct the pursuit of more focused studies on the mode of action thereof.

Biological Control And Endophytes

Biological control is the process whereby the population or vigour of one organism is limited when a second organism is placed within its ecologic sphere of influence. The application of parasites to control agricultural or other pests has guided use of this term for the most part. This is best demonstrated by the fact that the term "biological control agent" is replaced by the term "pathogen" in general usage where the organism controlled is a desirable one, although the biological interactions may be the same in each case.

In vegetation management, biological control of unwanted vegetation is ordinarily achieved by the introduction or application of living agents such as pathogenic microorganisms or herbivorous animals (e.g., bacteria, insects, sheep and goats) to target weeds. "Classical Biocontrol" involves the use of imported organisms as biocontrol agents and these are ordinarily used to control imported weeds (9). Indigenous fungi (Mycoherbicides) are generally used in the "Innundative Biocontrol" strategy (10). The latter technique involves application of inoculum of native pathogens to target pests in very large quantities.

By contrast, biological control, as applied to endophytes, has to date mostly emphasized resident or mutualistic fungi of grasses which render those hosts unpalatable to herbivores (4,5). Resident fungi of *Pseudotsuga menziesii* (Mirb.) Franco were also shown to render Douglas-fir tissues unpalatable to insects in several rather elegant experiments, described by Carroll (3). Another example was described by Petrini *et al.* (14) with respect to gall midge (*Paradiplosis tumifex* Gagné) of *Abies balsamea* (L.) Mill. Those endophytic fungi which cause feeding deterrence fit partially if not completely within the definition of mutualistic endophytes as put forward by Todd (20).

Endophytic fungi may also be used as biocontrol agents in what we consider to be a novel approach: the constraint of their host. It is our goal to promote the internal fungi of *Alnus rubra* Bong. from resident to necrotrophic status, whether by stimulating the fungi themselves, by reducing the physiologic vigour of the host plant or by achieving any suitable combination of the two.

We are working in British Columbia with some fungi that are not truly mutualistic, although they fit within the general broad definition of endophytes so long as the hosts remain healthy. Rather, they may in fact be quiescent pathogens, though their precise status within the apparently healthy plant is as yet undefined. Most important from the standpoint of research application in forestry, these are the fungi which are most often associated with premature death of young woody plants, such as *A. rubra*. Mortality of young (< 10 years) *A. rubra* in excess of 90% of local stands has been noted in the Pacific Northwest, often in association with numerous fructifications of *Melanconis alni* Tul. and *M. marginalis* (Peck) Wehm. (17). Although *M. alni* is regarded as a pathogen (8), its mode of infection and its precise status within the host tree are not defined. The same situation exists with *M. alni* as exists with many species of *Phomopsis*, *Fusarium*, and other fungi which are usually noted to be "associated" with this or that disease (12). Separation of primary pathogens, secondary pathogens and saprophytes becomes difficult when the living host tissues have been stressed in advance of infection and colonization.

Manipulation of Plants to Induce the Conversion of Mutualistic Symbionts to Pathogens

The conversion of mutualistic symbionts and latent endophytes generally to pathogens usually leads, in the present context, to the generation of a disease syndrome in the plant host. Symptom expression and/or fructification by the endophyte is most readily elicited by application of stress to the host plant. The chemical herbicide Vision® is used in most of the chemical brush clearance work in British Columbia. After application of Vision® to *A. rubra*, and sometimes prior to complete defoliation, fruiting bodies of *Melanconis* spp. become erumpent from the bark of the trees. Boddy & Griffith (1) were able to elicit fruiting by several fungi on hardwoods by slow-drying of stem sections. We were able to accomplish the same result with similar methods in British Columbia with *A. rubra* (unpublished). Fungi that occasionally developed from the bark of apparently healthy *A. rubra* stem sections subjected to slow-drying included *M. alni* and *M. marginalis*, two species which are most commonly collected in diseased natural stands of *A. rubra* in B.C. (17). However, the predominant fungus on *A. rubra* treated in this way was *Melanconis thelebola* (Fr.) Sacc., a species rarely seen fruiting on dying *A. rubra* in nature.

Sieber *et al.* (16), in analysis of experiments with endophytes of *Triticum aestivum* L., concluded: "Our results suggest that endophytic fungi could represent an additional factor which can reduce yield by consuming energy which is required for the defense response." If this proves to be the case generally, endophytes themselves might predispose their hosts to enviromental damage by reducing the damage threshold. Consequently, the complexity of host predisposition becomes a great deal more intricate. The significance of the presence of *M. thelebola* on *A. rubra* is under investigation on the basis of these considerations.

Manipulation of conditions affecting endophytic fungi in order to utilize their potential as necrotrophs will most often involve manipulation of the host. Two primary approaches are envisioned and are under consideration at this time:

1. The target host plant can be used as a conduit to translocate agents that stimulate the endophyte into necrotrophic activity, and

2. The target host plant is subjected to various stress agents, including the applications of topical chemicals and physical influences such as cold, heat, drying, etc., which alter the host-endophyte balance in favour of the endophyte.

In the first instance, chemostimulants of various types are available. Initial research will involve simply mass screening in order to specify an appropriate library of compounds which are both active and are translocated to the sites of activity.

In the second instance, application of stress (for example) which effectively limits water transport in the target plants will approximate slow-drying (1), to one degree or another. Additional stress agents will yield alternative types of host damage or debilitation. Thereafter, research can be directed toward collating and analyzing results achieved by methods which are functional through different modes of action.

The work envisioned is expensive and requires a group of trained scientific and technical personnel who are unavailable at the moment in our group. We are prepared to cooperate elsewhere in combined research efforts in order to gain maximum value from the resource of our institution and others directed toward solution of this problem.

Acknowledgements

The authors are grateful to Dr. Alvin Funk, T.A.D. Woods and Suzanne Wilson for assistance at various stages of this work which was supported in part by a grant from the Science and Technology Opportunities Fund, Forestry Canada, Ottawa.

Literature Cited

1. Boddy, L. and G.S. Griffith. 1989. Role of endophytes and latent invasion in the development of decay communities in sapwood of angiospermous trees. Sydowia 41:41-73.
2. Brunet, R. 1991. Herbicides: fighting back. Silviculture, Journal of the New Forest 6:10-11.
3. Carroll, G. 1988. Fungal endophytes in stems and leaves: from latent pathogen to mutualistic symbiont. Ecology 69:2-9.
4. Clay, K. 1988. Fungal endophytes of grasses: a defensive mutualism between plants and fungi. Ecology 69:10-16.
5. Clay, K. 1989. Clavicipitaceous endophytes of grasses: their potential as biocontrol agents. Mycol. Res. 92:1-12.
6. Comeau, P. and J. Boateng. 1992. Forest vegetation management in British Columbia. In: Proc. of Biocontrol of Forest Weeds - Proc. of a Workshop held at the Western International Forest Disease Work Conference in Vernon, B.C., Aug. 9, 1991, Pac. For. Centre, Forestry Canada, Victoria, B.C., 55pp.
7. Dorworth, C.E. 1990. Mycoherbicides for forest weed biocontrol - the P.F.C. enhancement process. Pages 116-119 in: Alternatives to the Chemical Control of Weeds, C. Bassett, L. J. Whitehouse, and J. A. Zabkiewicz, eds. FRI Bull. 155, Rotorua, New Zealand.
8. Funk, A. 1981. Parasitic microfungi of western trees. Pacific Forestry Centre Pub. No. BC-X-222, Victoria, B.C., 190 pp.
9. Harris, P. 1991. Classical biocontrol of weeds:Its definition, selection of effective agents, and administrative-political problems. Can. Ent. 123:827-849.
10. Hasan, S. and P.G. Ayres. 1990. The control of weeds through fungi: principles and prospects. New Phytol. 115:201-222.
11. Hawksworth, D.L., B.C. Sutton, and G.C. Ainsworth. 1983. Ainsworth and Bisby's Dictionary of the Fungi, 7th Ed., Commonwealth Mycol. Inst., Kew, U.K., 445 pp.

12. Hepting, G.H. 1971. Diseases of Forest and Shade Trees of the United States. U.S.D.A. - For. Serv. Hndbk. No. 386, Washington, DC. 658 pp.
13. Makowski, R. M. D. and K. Mortensen. 1992. The first mycoherbicide in Canada: *Colletotrichum gloeosporioides* f. sp. *malvae* for round-leaved mallow control. Pages 298-300 in: Proc. First Int. Weed Control Congr., Vol. 2, Melbourne, Australia, 17-21 Feb., 1992. (R.G. Richardson, Compiler), Weed Sci. Soc. Victoria Inc., Melbourne..
14. Petrini, L.E., O. Petrini and G. Laflamme. 1989. Recovery of endophytes of *Abies balsamea* from needles and galls of *Paradiplosis tumifex*. Phytoprotection 70:97-103.
15. Schoenweiss, D.F. 1975. Predisposition, stress and plant disease. Annu. Rev. Phytopathol. 13:193-211.
16. Sieber, T.N., T.K. Riesen, E. Müller, and P.M. Fried. 1988. Endophytic fungi in four winter wheat cultivars (*Triticum aestivum* L.) differing in resistance against *Stagonospora nodorum* (Berk.) Cast. & Germ. = *Septoria nodorum* (Berk.) Berk. J. Phytopathol. 122:289-306.
17. Sieber, T.N., F. Sieber-Canavesi, and C.E. Dorworth. 1991. Endophytic fungi of red alder (*Alnus rubra*) leaves and twigs in British Columbia. Can. J. Bot. 69:407-411.
18. Sieber, T.N., F. Sieber-Canavesi, O. Petrini, A.K.M. Ekramoddoullah, and C.E. Dorworth. 1991. Characterization of Canadian and European *Melanconium* from some *Alnus* species by morphological, cultural and biochemical studies. Can. J. Bot. 69:2170-2176.
19. Templeton, G.E. 1982. Biological herbicides: discovery, development, deployment. Weed Sci. 30:430-433.
20. Todd, D. 1988. The effects of host genotype, growth rate, and needle age on the distribution of a mutualistic, endophytic fungus in Douglas-fir plantations. Can. J. For. Res. 18:601-605.

INDEX